INTRODUCTION

Distinctive Home Designs

Ranch Home Plans is a collection of best-selling one-story plans from some of the nation's leading designers and architects. Only quality plans with sound design, functional layout, energy effeciency and affordability have been selected.

This plan book covers a wide range of architectural styles in a popular range of sizes. A broad assortment is presented to match a wide variety of lifestyles and budgets. Each design page features floor plans, a front view of the house, and a list of features. All floor plans show room dimensions, exterior dimensions and the interior square footage of the home.

Technical Specifications

Every effort has been made to ensure that these plans and specifications meet most nationally recognized building codes (BOCA, Southern Building Code Congress and others). Drawing modifications and/or the assistance of a local architect or professional designer are sometimes necessary to comply with local codes or to accommodate specific building site conditions.

Detailed Material Lists

An accurate material list showing the quantity, dimensions and description of the major building materials necessary to construct your new home can save you a considerable amount of time and money. See Home Plans Index on page 82 for availability.

Fax-A-Plan™

This is an ideal option for those who have interest in several home designs and want more information. Rear and side views for most designs in this publication are available via fax, along with a list of key construction features (i.e. roof slopes, ceiling heights, insulation values, type of roof and wall construction) and more. Just call our automated *FAX-A-PLAN* service at - 1-314-770-2228 available 24 hours a day - 7 days a week. Use of this service is free of charge.

Blueprint Ordering - Fast and Easy

Your ordering is made simple by following the instructions on page 83. See page 82 for more information on what type of blueprint packages are available and how many plan sets to order.

Your Home, Your Way

The blueprints you receive are a master plan for building your new home. They start you on your way to what may well be the most rewarding experience of your life.

CONTENTS

House shown on front cover is Plan #554-0706 and is featured on page 56.

Ranch Home Plans is published by Home Design Alternatives, Inc. (HDA, Inc.) 4390 Green Ash Drive, St. Louis, MO 63045. All rights reserved. Reproduction in whole or in part without written permission of the publisher is prohibited. Printed in U.S.A © 2001. Artist drawings shown in this publication may vary slightly from the actual working blueprints.

Copyright All plans appearing in this publication are protected under copyright law. Reproduction of the illustrations or working drawings by any means is strictly prohibited. The right of building only one structure from the plans purchased is licensed exclusively to the buyer and the plans may not be resold unless by express written authorization.

HOW YOU CAN CUSTOMIZE
OUR PLANS INTO YOUR DREAM HOME

Many of the plans in this book are customizable through the use of our exclusive Customizer Kit™. Look for this symbol on the plan pages for availability.

With the Customizer Kit you have unlimited design possibilities available to you when building a new home. It allows you to alter virtually any architectural element you wish, both on the exterior and interior of the home. The Kit, available with many of the home plans, allows you to alter virtually any architectural element you wish, both on the exterior and interior of the home. The Kit comes complete with simplified drawings of your selected home plan so that you can sketch out any and all of your changes. To help you through this process, the Kit also includes a workbook called "The Customizer," a special correction pen, a red marking pencil, an architect's scale and furniture layout guides. These tools, along with the simplified customizer drawings, allow you to experiment with various design changes prior to having a design professional modify the actual working drawings.

Before placing your order for blueprints consider the type and number of changes you plan to make to your selected design. If you wish to make only minor design changes such as moving interior walls, changing window styles, or altering foundation types, we strongly recommend that you purchase reproducible masters along with the

Customizer Kit. These master drawings, which contain the same information as the blueprints, are easy to modify because they are printed on erasable, reproducible paper. Also, by starting with complete detailed drawings, and planning out your changes with the Customizer Kit, the cost of having a design professional or your builder make the required drawing changes will be considerably less. After the master drawings are altered, multiple blueprint copies can be made from them.

If you anticipate making a lot of changes, such as moving exterior walls and changing the overall appearance of the house, we suggest you purchase only one set of blueprints as a reference set and the Customizer Kit to document your desired changes. When making major design changes, it is always advisable to seek out the

Figure 2

assistance of an architect or design professional to review and redraw that portion of the blueprints affected by your changes.

Typically, having a set of reproducible masters altered by a local designer can cost as little as a couple hundred dollars, whereas redrawing a portion or all of the blueprints can cost considerably more depending on the extent of the changes. Like most projects, the more planning and preparation you can do on your own, the greater the savings to you.

Finally, you'll have the satisfaction of knowing that your custom home is uniquely and exclusively yours.

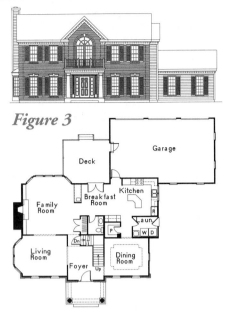

Figure 3

Figure 1

EXAMPLES OF CUSTOMIZING

Thousands of builders and home buyers have used the HDA Customizer Kit to help them modify their home plans, some involving minor changes, many with dramatic alterations. Examples of actual projects are shown here.

Figure 1 shows the front elevation and first floor plan for one of HDA's best-selling designs.

Figure 2 shows how one plan customer made few but important design changes such as completely reversing the plan to better accommodate his building site; adding a second entrance for ease of access to the front yard from the kitchen; making provisions for a future room over the garage by allowing for a stairway and specifying windows in place of louvers, plus other modifications.

Figure 3 shows another example of an actual project where the design shown in Figure 1 was dramatically changed to achieve all of the desired features requested by the customer. This customized design proved to be so successful that HDA obtained permission to offer it as a standard plan.

Double Atrium Embraces The Sun

3,199 total square feet of living area

Special features

- Grand scale kitchen features bay-shaped cabinetry built over atrium that overlooks two-story window wall

- A second atrium dominates the master suite which boasts a sitting area with bay window and luxurious bath, which has whirlpool tub open to the garden atrium and lower level study

- 3 bedrooms, 2 1/2 baths, 3-car side entry garage

- Walk-out basement foundation

- 2,349 square feet on the first floor and 850 square feet on the lower level

Price Code E

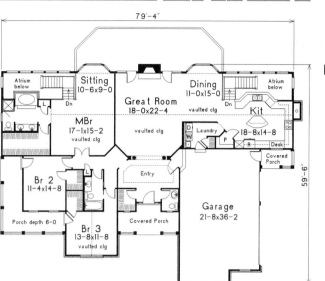

First Floor

79'-4"

Atrium below | Sitting 10-6x9-0 | Dining 11-0x15-0 | Atrium below
Great Room 18-0x22-4 vaulted clg
MBr 17-1x15-2 vaulted clg | Kit 18-8x14-8
Laundry | Desk | Covered Porch
Br 2 11-4x14-8 | Entry | Garage 21-8x36-2
Porch depth 6-0 | Covered Porch
Br 3 13-8x11-8 vaulted clg
59'-6"

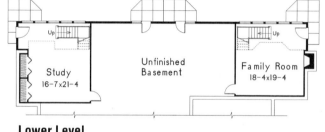

Lower Level

Study 16-7x21-4 | Unfinished Basement | Family Room 18-4x19-4

Customize This Plan **SEE PAGE 4**

Material List Available **SEE PAGE 81**

Rear View

SUMMERVIEW

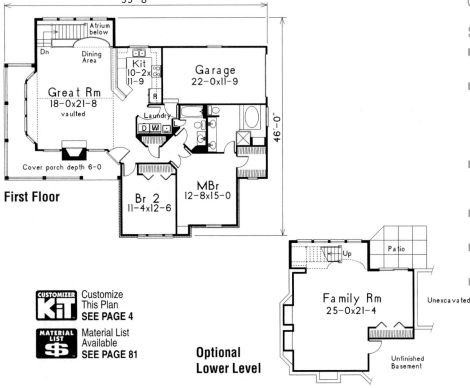

Rear View

Tranquility Of An Atrium Cottage

1,384 total square feet of living area

Special features

- Wrap-around country porch for peaceful evenings
- Vaulted great room enjoys a large bay window, stone fireplace, pass-through kitchen and awesome rear views through atrium window wall
- Master suite features double entry doors, walk-in closet and a fabulous bath
- Atrium open to 611 square feet of optional living area below
- 2 bedroom, 2 baths, 1-car side entry garage
- Walk-out basement foundation

Price Code A

First Floor

55'-8"

46'-0"

Atrium below

Dn

Dining Area

Kit 10-2x 11-9

Garage 22-0x11-9

Great Rm 18-0x21-8 vaulted

Laundry

D W

R

Cover porch depth 6-0

Br 2 11-4x12-6

MBr 12-8x15-0

CUSTOMIZER KiT — Customize This Plan **SEE PAGE 4**

MATERIAL LIST $ — Material List Available **SEE PAGE 81**

Optional Lower Level

Up

Patio

Family Rm 25-0x21-4

Unexcavated

Unfinished Basement

Plan #554-0732

Country Home With Front Orientation

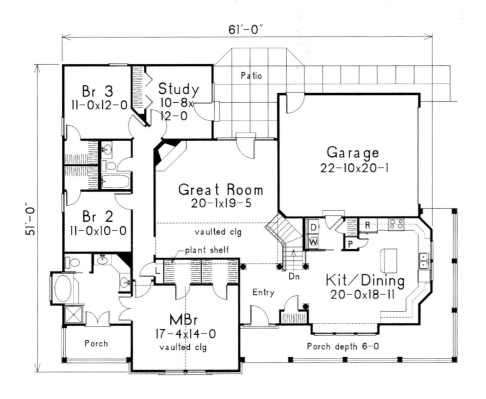

2,029 total square feet of living area

Special features

- Stonework, gables, roof dormer and double porches create a country flavor

- Kitchen enjoys extravagant cabinetry and counterspace in a bay, island snack bar, built-in pantry and cheery dining area with multiple tall windows

- Angled stair descends from large entry with wood columns and is open to vaulted great room with corner fireplace

- Master bedroom boasts his and hers walk-in closets, double-doors leading to an opulent master bath and private porch

- 3 bedrooms, 2 baths, 2-car side entry garage

- Basement foundation

  Price Code C

CUSTOMIZER KiT — Customize This Plan **SEE PAGE 4**

MATERIAL LIST $ — Material List Available **SEE PAGE 81**

CARLSTON

Classic Elegance

2,483 total square feet of living area

Special features

- A large entry porch with open brick arches and palladian door welcomes guests

- The vaulted great room features an entertainment center alcove and ideal layout for furniture placement

- Dining room is extra large with a stylish tray ceiling

- 3 bedrooms, 2 baths, 2-car side entry garage

- Basement foundation

  Price Code D

 Customize This Plan **SEE PAGE 4** Material List Available **SEE PAGE 81**

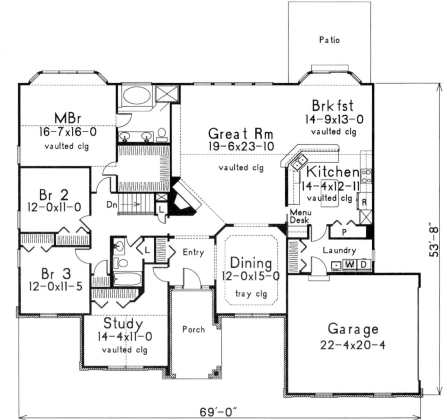

Plan #554-0719

Classic Atrium Ranch With Rooms To Spare

1,977 total square feet of living area

Special features

- Classic traditional exterior always in style
- Spacious great room boasts a vaulted ceiling, dining area, atrium with elegant staircase and feature windows
- Atrium open to 1,416 square feet of optional living area below and consists of an optional family room, two bedrooms, two baths and a study
- 4 bedrooms, 2 1/2 baths, 3-car side entry garage
- Walk-out basement foundation
- 1,977 square feet on the first floor and 1,416 optional square feet on the lower level

Price Code C

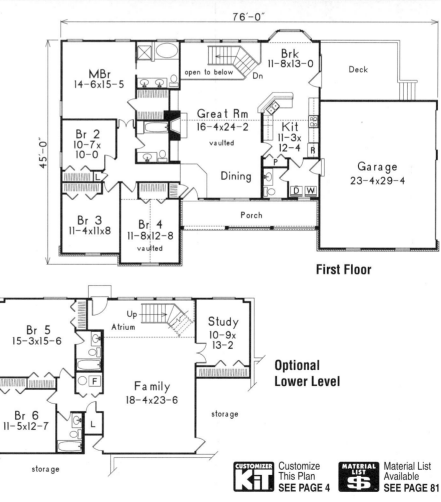

First Floor

Optional Lower Level

CUSTOMIZER KiT — Customize This Plan SEE PAGE 4

MATERIAL LIST $ — Material List Available SEE PAGE 81

RYLAND

Classic Ranch Has Grand Appeal With Expansive Porch

> 1,400 total square feet of living area

Special features

- Master bedroom is secluded for privacy
- Large utility room with additional cabinet space
- Covered porch provides an outdoor seating area
- Roof dormers add great curb appeal
- Vaulted ceilings in living room and master bedroom
- Oversized two-car garage with storage
- 3 bedrooms, 2 baths, 2-car garage
- Basement foundation, drawings also include crawl space foundation

        Price Code A

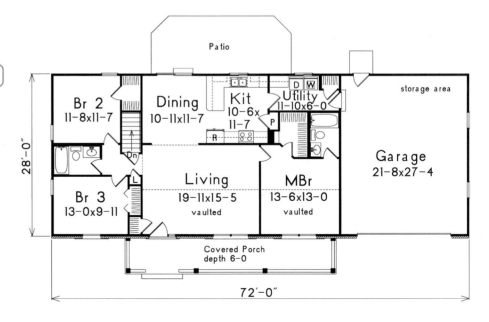

CUSTOMIZER KiT Customize This Plan **SEE PAGE 4**

MATERIAL LIST $ Material List Available **SEE PAGE 81**

Plan #554-0690

Atrium's Dramatic Ambiance, Compliments Of Windows

1,721 total square feet of living area

Special features

- Roof dormers add great curb appeal
- Vaulted great room and dining room immersed in light from atrium window wall
- Breakfast room opens onto covered porch
- Functionally designed kitchen
- 3 bedrooms, 2 baths, 3-car garage
- Walk-out basement foundation, drawings also include crawl space and slab foundations

Price Code C

CUSTOMIZER KiT — Customize This Plan **SEE PAGE 4**

MATERIAL LIST $$ — Material List Available **SEE PAGE 81**

Rear View

Plan #554-0370

VALRICO

Impressive Master Suite

2,287 total square feet of living area

Special features

- Double-doors lead into an impressive master suite which accesses covered porch and features deluxe bath with double closets and step-up tub
- Kitchen easily serves formal and informal areas of home
- The spacious foyer opens into formal dining and living rooms
- 4 bedrooms, 2 1/2 baths, 2-car side entry garage
- Slab foundation

Price Code E

CUSTOMIZER KIT — Customize This Plan **SEE PAGE 4**

MATERIAL LIST $ — Material List Available **SEE PAGE 81**

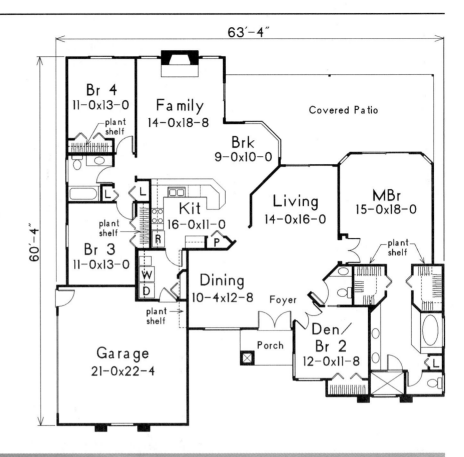

63'-4"

60'-4"

Br 4
11-0x13-0

plant shelf

Family
14-0x18-8

Covered Patio

Brk
9-0x10-0

Living
14-0x16-0

MBr
15-0x18-0

plant shelf

Kit
16-0x11-0

plant shelf

Br 3
11-0x13-0

Dining
10-4x12-8

Foyer

Porch

Den/
Br 2
12-0x11-8

plant shelf

Garage
21-0x22-4

Plan #554-0339

Plan #554-ES-103-1 basement
Plan #554-ES-103-2 crawl space & slab

Multi-Roof Levels Create Attractive Colonial Home

1,364 total square feet of living area

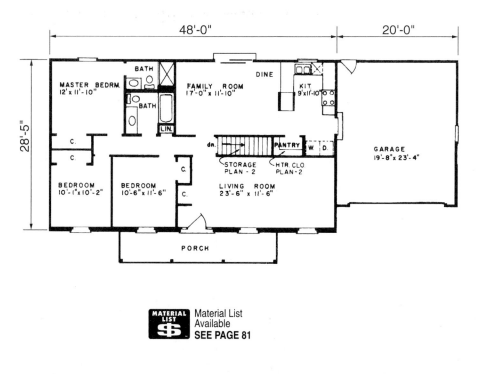

MATERIAL LIST $ Material List Available SEE PAGE 81

Special features

- A large porch and entry door with sidelights lead into a generous living room
- Well-planned U-shaped kitchen features a laundry closet, built-in pantry and open peninsula
- Master bedroom has its own bath with 4' shower
- Convenient to the kitchen is an oversized two-car garage with service door to rear
- 3 bedrooms, 2 baths, 2-car garage
- Basement, crawl space or slab foundation available, please specify when ordering

Price Code A

STRATFORD

Vaulted Ceilings And Light Add Dimension

1,676 total square feet of living area

Special features

- The living area skylights and large breakfast room with bay window provide plenty of sunlight

- The master bedroom has a walk-in closet and both the secondary bedrooms have large closets

- Vaulted ceilings, plant shelving and a fireplace provide a quality living area

- 3 bedrooms, 2 baths, 2-car garage

- Basement foundation, drawings also include crawl space and slab foundations

Price Code B

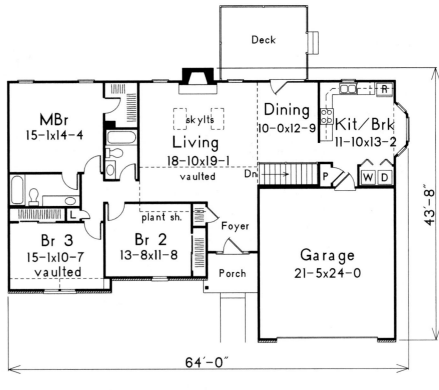

Deck

MBr
15-1x14-4

skylts

Living
18-10x19-1
vaulted

Dining
10-0x12-9

Kit/Brk
11-10x13-2

Dn

Br 3
15-1x10-7
vaulted

plant sh.

Br 2
13-8x11-8

Foyer

Porch

Garage
21-5x24-0

43'-8"

64'-0"

CUSTOMIZER KiT Customize This Plan **SEE PAGE 4**

MATERIAL LIST $ Material List Available **SEE PAGE 81**

Plan #554-0229

Private Breakfast Room Provides Casual Dining

1,708 total square feet of living area

Special features

- Massive family room enhanced with several windows, fireplace and access to porch
- Deluxe master bath accented by step-up corner tub flanked by double vanities
- Closets throughout maintain organized living
- Bedrooms isolated from living areas
- 3 bedrooms, 2 baths, 2-car garage
- Basement foundation, drawings also include crawl space foundation

    Price Code B

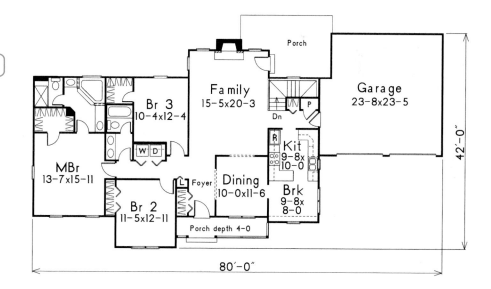

Customize This Plan SEE PAGE 4

Material List Available SEE PAGE 81

Plan #554-0450

IMPERSA

Great Expectations

2,190 total square feet of living area

Special features

- Comfort and good looks are combined in this design for modern families
- Perfect for those who love a sense of indoor/outdoor living
- Enter pillared front entry to the foyer with built-in planter
- Rear of home features centrally located kitchen and garden/ breakfast area with access to optional rear deck with raised planter or optional hot tub
- 3 bedrooms, 2 baths, 2-car garage
- Basement foundation, drawings also include slab foundation

Price Code C

Material List Available
SEE PAGE 81

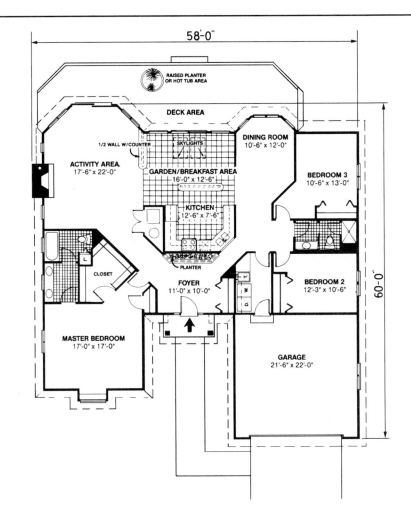

Plan #554-1260

Traditional Exterior, Handsome Accents

1,882 total square feet of living area

Special features

- Wide, handsome entrance opens to the vaulted great room with fireplace
- Living and dining areas are conveniently joined but still allow privacy
- Private covered porch extends breakfast area
- Practical passageway runs through laundry and mud room from garage to kitchen
- Vaulted ceiling in master bedroom
- 3 bedrooms, 2 baths, 2-car garage
- Basement foundation
    Price Code D

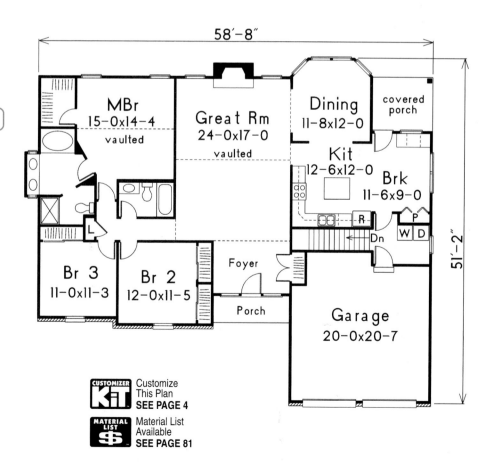

CUSTOMIZER KiT Customize This Plan **SEE PAGE 4**

MATERIAL LIST $$ Material List Available **SEE PAGE 81**

BRIARWOOD

Enchanting Country Cottage

1,140 total square feet of living area

Special features

- Open and spacious living and dining area for family gatherings
- Well-organized kitchen with an abundance of cabinetry and built-in pantry
- Roomy master bath features double-bowl vanity
- 3 bedrooms, 2 baths, 2-car drive under garage
- Basement foundation

Price Code AA

CUSTOMIZER KiT Customize This Plan SEE PAGE 4

MATERIAL LIST $ Material List Available SEE PAGE 81

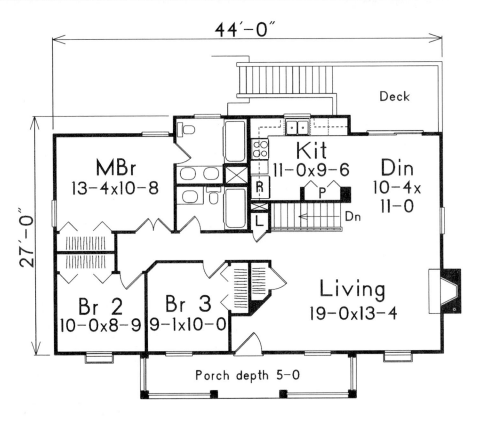

Deck

44´-0˝

27´-0˝

MBr
13-4x10-8

Kit
11-0x9-6

Din
10-4x 11-0

R
P
L
Dn

Br 2
10-0x8-9

Br 3
9-1x10-0

Living
19-0x13-4

Porch depth 5-0

Plan #554-0477

Grand Entryway Adorns This Home

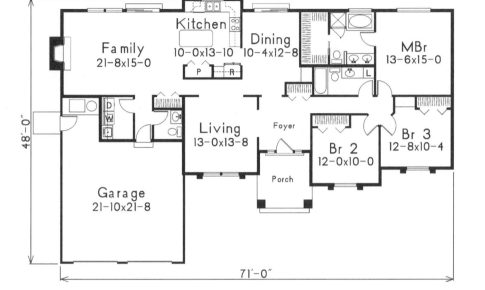

Family 21-8x15-0

Kitchen 10-0x13-10

Dining 10-4x12-8

MBr 13-6x15-0

P R

Living 13-0x13-8

Foyer

Br 3 12-8x10-4

Br 2 12-0x10-0

Porch

Garage 21-10x21-8

48'-0"

71'-0"

CUSTOMIZER KIT Customize This Plan **SEE PAGE 4**

MATERIAL LIST $ Material List Available **SEE PAGE 81**

| 1,941 total square feet of living area |

Special features

- Kitchen incorporates a cooktop island, a handy pantry and adjoins the dining and family rooms
- Formal living room, to the left of the foyer, lends a touch of privacy
- Raised ceiling in foyer, living, dining and kitchen areas
- Laundry room, half bath and closet all located near the garage
- Both the dining and family rooms have access outdoors through sliding doors
- 3 bedrooms, 2 1/2 baths, 2-car garage
- Crawl space foundation

  Price Code C

Plan #554-0682

APOLLO

Plan #554-1279-1 partial basement/crawl space
Plan #554-1279-2 slab

Attractive Arched Main Entrance

| 2,136 total square feet of living area |

Special features

- Vaulted breakfast nook includes bay window
- Kitchen is tucked away and includes sloped ceiling and built-in pantry
- Large deck off activity area for expanding your entertaining options
- Master bedroom with walk-in closet and private bath
- 3 bedrooms, 2 baths, 2-car garage
- Partial basement/crawl space or slab foundation available, please specify when ordering

Price Code C

Material List Available
SEE PAGE 81

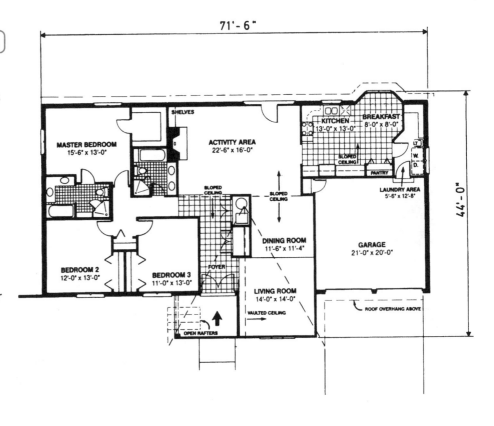

Convenient Ranch

1,120 total square feet of living area

Special features

- Master bedroom includes a half bath with laundry area, linen closet and kitchen access
- Kitchen has charming double-door entry, breakfast bar and a convenient walk-in pantry
- Welcoming front porch opens to large living room with coat closet
- 3 bedrooms, 1 1/2 baths
- Crawl space foundation, drawings also include basement and slab foundations

Price Code AA

40'-0"

28'-0"

W D

MBr
10-0x11-8

L

L P
S R

Kit
8-1x
13-0

Dining
10-0x
13-0

Br 2
10-0x
10-8

Br 3
9-0x
10-8

Living
17-5x14-1

Porch depth 4-0

CUSTOMIZER **KiT** Customize This Plan **SEE PAGE 4**

MATERIAL LIST **$** Material List Available **SEE PAGE 81**

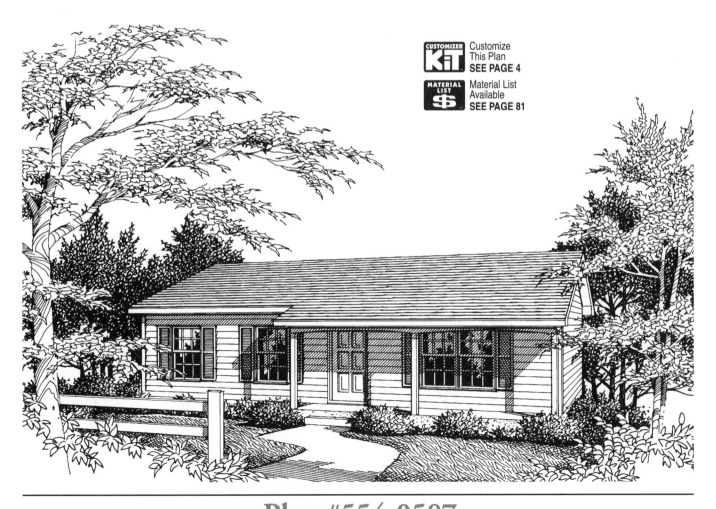

WILSHIRE

Efficient And Attractive

2,076 total square feet of living area

Special features

- Superbly designed kitchen will make food preparation a breeze

- Family room and patio are perfectly located for side yard oriented views

- Large living and dining rooms capitalize on rear views while still convenient to entry

- A compartmented bath and good-sized walk-in closet adjoin the master bedroom

- 4 bedrooms, 2 baths, 2-car side entry garage

- Partial basement/crawl space or crawl space foundation available, please specify when ordering

Price Code C

Material List Available **SEE PAGE 81**

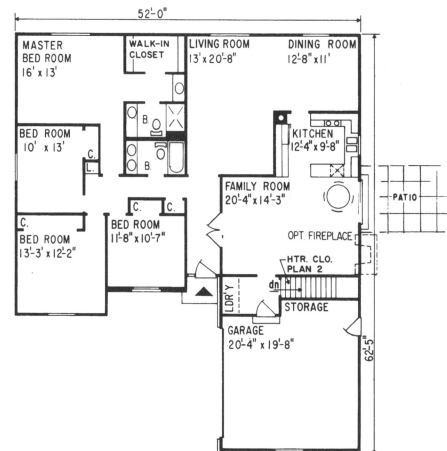

Plan #554-ES-124-1 partial basement/crawl space
Plan #554-ES-124-2 crawl space

Plan #554-ES-124-1 & 2

MATERIAL LIST $$

Material List
Available
SEE PAGE 81

Plan #554-1123-1 basement
Plan #554-1123-2 crawl space & slab

Efficient Use Of Space

> 1,513 total square feet of living area

Special features

- Master bedroom features private bath

- Nice-sized U-shaped kitchen overlooks large eating area and is completely open to family room with optional fireplace

- Side entry into home is convenient to garage, basement and family room

- First floor laundry closet adjoins breakfast area

- 3 bedrooms, 2 baths, 2-car garage

- Basement, crawl space or slab foundation available, please specify when ordering

Price Code B

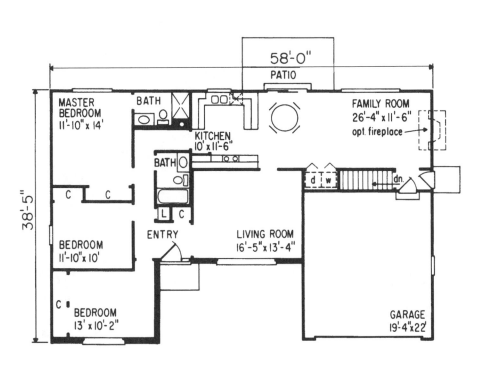

DELMAR

Central Fireplace Focuses Family Living

1,408 total square feet of living area

Special features

- Handsome see-through fireplace offers a gathering point for the family room and breakfast/kitchen area

- Vaulted ceiling and large bay window in the master bedroom add charm to this room

- A dramatic angular wall and large windows add brightness to the kitchen/breakfast area

- Family room and breakfast/kitchen area have vaulted ceilings, adding to this central living area

- 3 bedrooms, 2 baths, 2-car garage

- Crawl space foundation, drawings also include slab foundation

Price Code A

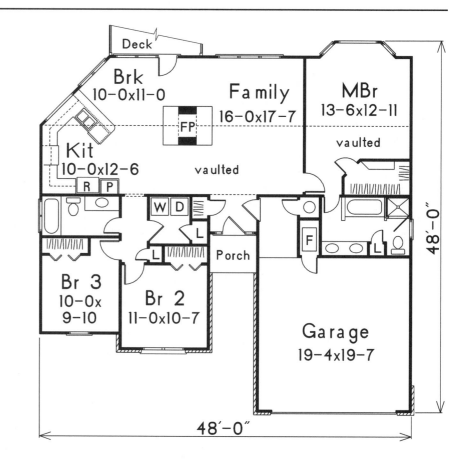

Customize This Plan **SEE PAGE 4**

Material List Available **SEE PAGE 81**

SPARROW

Plan #554-ES-107-1 basement
Plan #554-ES-107-2 crawl space & slab

Compact Ranch With Good Looks

1,120 total square feet of living area

Special features

- Porch and shuttered windows with lower accent panels add greatly to this home's appeal
- Kitchen offers snack counter and opens to family room
- All bedrooms provide excellent closet space
- Carport includes building for ample storage
- 3 bedrooms, 1 1/2 baths, 1-car carport
- Basement, crawl space or slab foundation available, please specify when ordering

    Price Code A

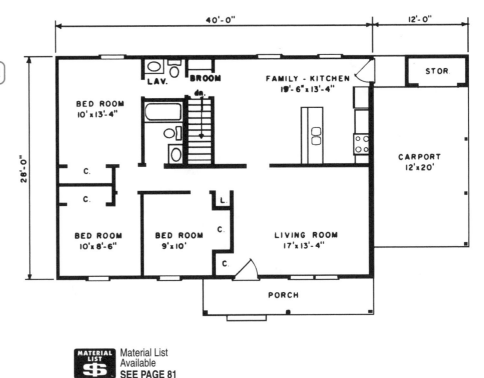

MATERIAL LIST
$ Material List Available
SEE PAGE 81

CARRMONT

Well-Designed Family Living

1,441 total square feet of living area

Special features

- Combined living/dining room offers plenty of open space for large family gatherings
- Functional kitchen opens to nice-sized family room with sliding glass doors opening to rear patio
- Master bedroom offers a private bath; two other bedrooms share a hall bath
- Convenient utility room opens to front foyer and garage
- 3 bedrooms, 2 baths, 2-car garage
- Basement, crawl space or slab foundation available, please specify when ordering

Price Code A

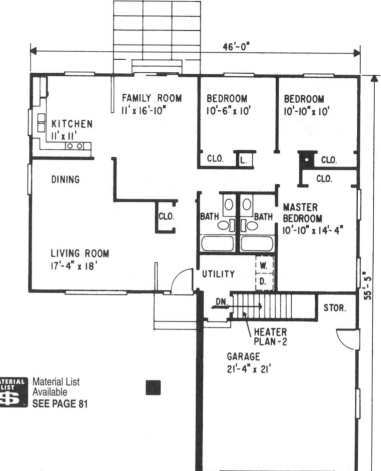

MATERIAL LIST $ Material List Available **SEE PAGE 81**

Plan #554-N292-1 basement
Plan #554-N292-2 crawl space & slab

Plan #554-N292-1 & 2

Plan #554-1243-1 partial basement/crawl space
Plan #554-1243-2 crawl space

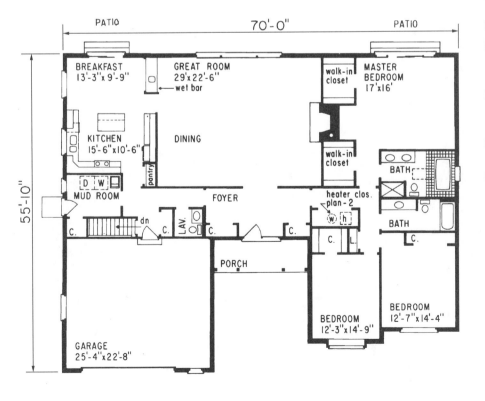

PATIO — 70'-0" — PATIO

BREAKFAST 13'-3" x 9'-9"

GREAT ROOM 29'x22'-6"
← wet bar

walk-in closet

MASTER BEDROOM 17'x16'

KITCHEN 15'-6"x10'-6"

DINING

walk-in closet

BATH

55'-10"

D W
MUD ROOM

pantry

FOYER

heater clos. plan-2
w   h

BATH

C.

dn

C. LAV. C.

C.

C. L

C.

PORCH

BEDROOM 12'-3"x14'-9"

BEDROOM 12'-7"x14'-4"

GARAGE 25'-4"x22'-8"

Charming Home With Pleasing Proportions

2,705 total square feet of living area

Special features
- Brick veneer creates a striking contrast with stucco and rough-sawn timber gables
- Covered front porch leads to foyer and great room at rear of home
- Great room includes wet bar and fireplace on interior wall
- Breakfast room, with access to patio, opens to kitchen with island work center and pantry
- 3 bedrooms, 2 1/2 baths, 2-car garage
- Partial basement/crawl space or crawl space foundation available, please specify when ordering

Price Code E

Material List Available
SEE PAGE 81

WOODLAWN

High Ceilings
Create Openness

CUSTOMIZER KiT — Customize This Plan **SEE PAGE 4**

MATERIAL LIST $ — Material List Available **SEE PAGE 81**

2,516 total square feet of living area

Special features

- 12' ceiling in living areas
- Plenty of closet space in this open ranch plan
- Large kitchen/breakfast area joins great room via see-through fireplace creating large entering space
- Large three-car garage has extra storage area
- Master bedroom has eye-catching bay window
- 3 bedroom, 2 1/2 baths, 3-car garage
- Basement foundation

    Price Code D

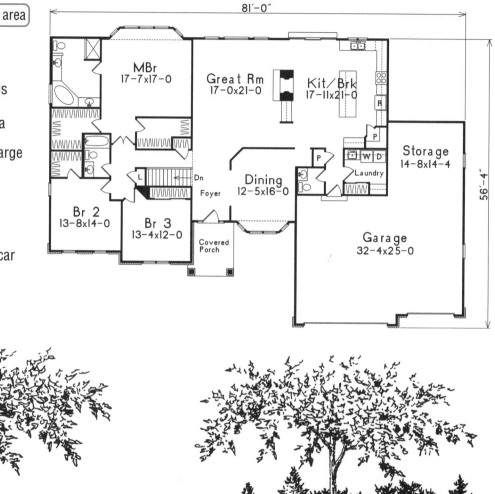

Plan #554-0746

To order blueprints use the form on page 83 or call **1-800-DREAM HOME** (373-2646)

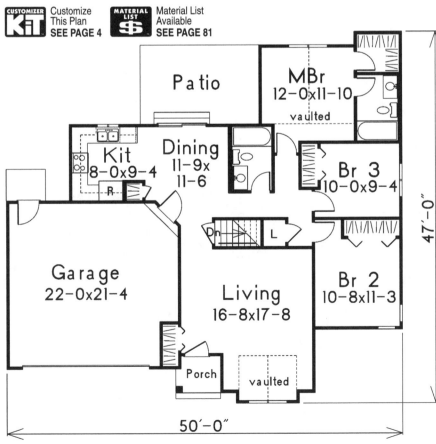

Patio

MBr
12-0x11-10
vaulted

Kit
8-0x9-4

Dining
11-9x
11-6

R

Dn

L

Br 3
10-0x9-4

Garage
22-0x21-4

Living
16-8x17-8

Br 2
10-8x11-3

Porch

vaulted

47'-0"

50'-0"

CUSTOMIZER KiT Customize This Plan **SEE PAGE 4**

MATERIAL LIST $ Material List Available **SEE PAGE 81**

Vaulted Ceiling Frames Circle-Top Window

1,195 total square feet of living area

Special features

- Kitchen/dining room opens onto the patio
- Master bedroom features vaulted ceiling, private bath and walk-in closet
- Coat closets located by both the entrances
- Convenient secondary entrance at the back of the garage
- 3 bedrooms, 2 baths, 2-car garage
- Basement foundation

  Price Code AA

CHESWICK

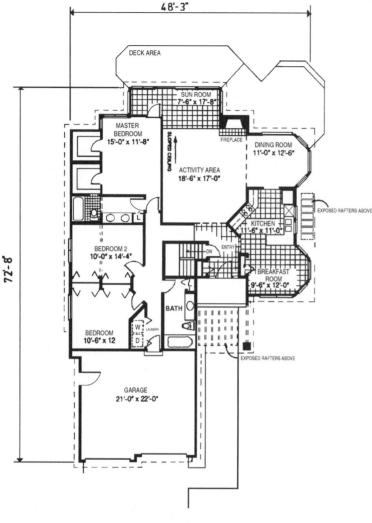

48'-3"

72'-8"

DECK AREA

SUN ROOM
7'-6" x 17'-8"

MASTER
BEDROOM
15'-0" x 11'-8"

SLOPED CEILING

FIREPLACE

DINING ROOM
11'-0" x 12'-6"

ACTIVITY AREA
18'-6" x 17'-0"

EXPOSED RAFTERS ABOVE

KITCHEN
11'-6" x 11'-0"

BEDROOM 2
10'-0" x 14'-4"

DN

ENTRY

BREAKFAST
ROOM
9'-6" x 12'-0"

BATH

W
D   LAUNDRY

BEDROOM
10'-6" x 12

EXPOSED RAFTERS ABOVE

GARAGE
21'-0" x 22'-0"

Handsome Contemporary Design Marks This Home As Special

1,907 total square feet of living area

Special features

- Entry foyer opens to kitchen and breakfast area on the right and a large activity area on the left

- Activity area amenities include a fireplace and sun room

- Formal dining area with bay windows and sliding glass doors located at rear of kitchen

- Master bedroom has his and hers walk-in closets and a dual-vanity bath

- Two additional bedrooms share one full bath

- 3 bedrooms, 2 baths, 2-car garage

- Partial basement/crawl space foundation

Price Code C

Plan #554-1324

OAKSHIRE

Customize This Plan **SEE PAGE 4**

Material List Available **SEE PAGE 81**

Contemporary Elegance With Efficiency

1,321 total square feet of living area

Special features

- Rear garage and elongated brick wall adds to appealing facade
- Dramatic vaulted living room includes corner fireplace and towering feature windows
- Kitchen/breakfast room is immersed in light from two large windows and glass sliding doors
- 3 bedrooms, 2 baths, 1-car rear entry garage
- Basement foundation

Price Code A

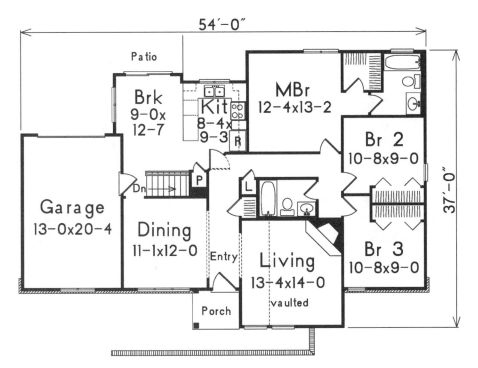

Plan #554-0660

To order blueprints use the form on page 83 or call **1-800-DREAM HOME** *(373-2646)*

31

BELLEAIR

Material List
Available
SEE PAGE 81

Affordable And Spaciously Designed Ranch

1,288 total square feet of living area

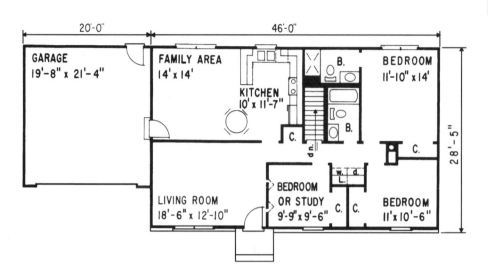

GARAGE
19'-8" x 21'-4"

20'-0"

46'-0"

FAMILY AREA
14' x 14'

KITCHEN
10' x 11'-7"

B.

BEDROOM
11'-10" x 14'

28'-5"

C.

LIVING ROOM
18'-6" x 12'-10"

BEDROOM
OR STUDY
9'-9"x 9'-6"

C. C.

BEDROOM
11'x10'-6"

Special features

- Large living room expands visually through bi-fold doors when third bedroom is utilized as a study
- Spacious eat-in kitchen adjoins sunny family room
- First floor laundry closet makes laundry easy
- Cheerful master bedroom has private bath and shower
- 3 bedrooms, 2 baths, 2-car garage
- Basement, crawl space or slab foundation available, please specify when ordering

    Price Code A

Plan #554-ES-144-1 basement
Plan #554-ES-144-2 crawl space & slab

Plan #554-ES-144-1 & 2

Plan #554-1261-1 partial basement/crawl space
Plan #554-1261-2 slab

Spanish Ranch

2,050 total square feet of living area

Special features

- Spanish facade with raised planters, spanish-clay roofing and exposed rafters make this ranch-style home extra charming
- Entrance leads to sloped ceiling, sunken living/dining room on right
- Activity area at the rear of the home includes a fireplace and snack bar
- U-shaped kitchen is accessed from the garage
- Left-wing sleeping quarters include a master bedroom with private bath featuring raised tub and shower and two additional bedrooms with a full bath
- 3 bedrooms, 2 1/2 baths, 2-car garage
- Partial basement/crawl space or slab foundation available, please specify when ordering

Price Code C

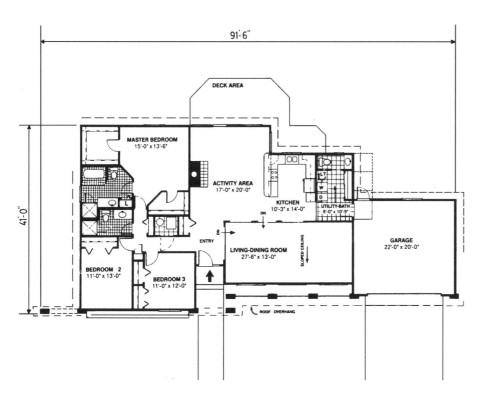

BELLPORT

Cozy Ranch Perfect For Starter Home

MATERIAL LIST $ Material List Available SEE PAGE 81

1,408 total square feet of living area

Special features

- Front entry has coat closet and opens to a view of the spacious living room
- Nice-sized family room offers an optional fireplace and sliding doors opening to rear patio
- Plenty of room for a workbench and storage in this oversized garage
- 3 bedrooms, 2 baths, 2-car garage
- Basement/crawl space, crawl space or slab foundation available, please specify when ordering

    Price Code A

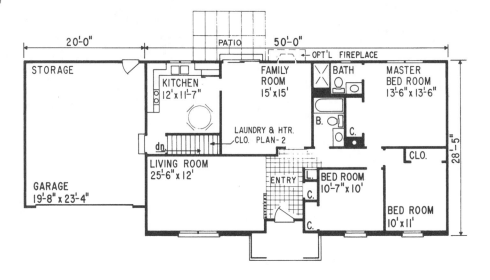

Plan #554-N286-1 basement
Plan #554-N286-2 crawl space & slab

Plan #554-N286-1 & 2

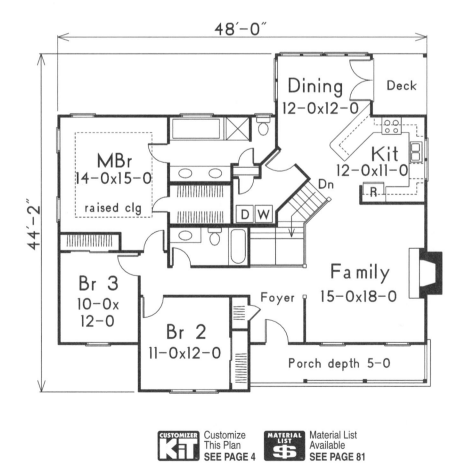

48´-0″

44´-2″

Dining
12-0x12-0

Deck

MBr
14-0x15-0

raised clg

D W

Dn

Kit
12-0x11-0

R

Family
15-0x18-0

Br 3
10-0x
12-0

Foyer

Br 2
11-0x12-0

Porch depth 5-0

Family Room With Fireplace Perfect For Central Gathering

[1,631 total square feet of living area]

Special features

- 9' ceilings throughout this home
- Utility room conveniently located near kitchen
- Roomy kitchen and dining areas boast a breakfast bar and patio access
- Coffered ceiling accents master suite
- 3 bedrooms, 2 baths, 2-car drive under garage
- Basement foundation

    Price Code B

CUSTOMIZER KiT Customize This Plan **SEE PAGE 4**

MATERIAL LIST $ Material List Available **SEE PAGE 81**

Plan #554-0237

OAKLAND

L-Shaped Kitchen Joined With Dining Area

CUSTOMIZER **KiT** — Customize This Plan **SEE PAGE 4**

MATERIAL LIST **$** — Material List Available **SEE PAGE 81**

1,000 total square feet of living area

Special features

- Master bedroom with double closets and adjacent bath
- L-shaped kitchen includes side entrance, closet and convenient laundry area
- Living room features handy coat closet
- 3 bedrooms, 1 bath
- Crawl space foundation, drawings also include basement and slab foundations

Price Code AA

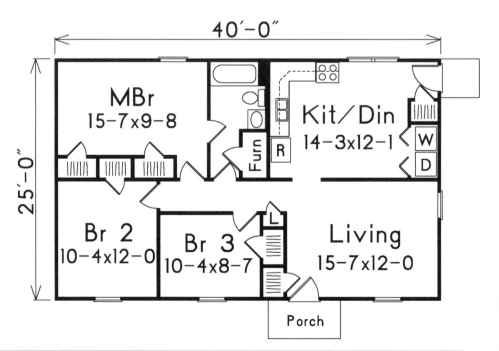

40'-0"

25'-0"

MBr 15-7x9-8

Fur

Kit/Din 14-3x12-1

W
D
R

Br 2 10-4x12-0

Br 3 10-4x8-7

L

Living 15-7x12-0

Porch

Plan #554-0583

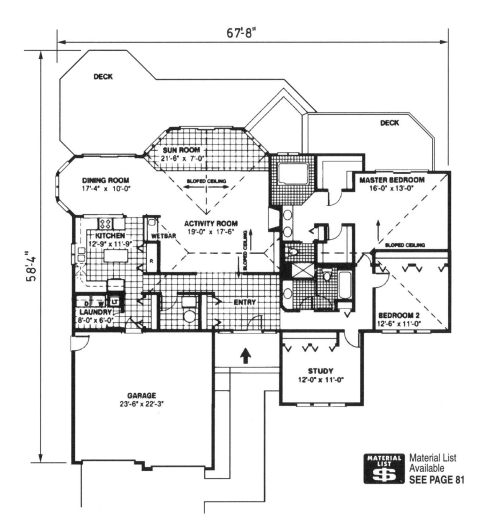

67'-8"

58'-4"

DECK

DECK

SUN ROOM
21'-6" x 7'-0"

SLOPED CEILING

DINING ROOM
17'-4" x 10'-0"

MASTER BEDROOM
16'-0" x 13'-0"

ACTIVITY ROOM
19'-0" x 17'-6"

SLOPED CEILING

KITCHEN
12'-9" x 11'-9"

WETBAR

SLOPED CEILING

P.

BEDROOM 2
12'-6" x 11'-0"

LAUNDRY
8'-0" x 6'-0"

ENTRY

GARAGE
23'-6" x 22'-3"

STUDY
12'-0" x 11'-0"

MATERIAL LIST $
Material List
Available
SEE PAGE 81

Sunny Retreat

2,180 total square feet of living area

Special features

- Exterior provides eye-catching roof lines
- Entry has cathedral ceiling and leads to activity room which features vaulted ceilings, sunken sun room and wet bar
- Master bedroom has dual walk-in closets, raised tub and compartmented shower
- Front-facing study with closet would make a perfect office or third bedroom
- 3 bedrooms, 2 baths, 2-car garage
- Crawl space foundation

    Price Code C

MOUNTAINVIEW

Begging For A Country Setting

1,293 total square feet of living area

Special features

- A very affordable ranch home that's easy to build
- Living room has separate entry, guest closet and opens to dining area
- Eat-in L-shaped kitchen offers pass-through to family room
- Master bedroom has its own bath and large walk-in closet
- 3 bedrooms, 2 baths, 1-car garage
- Basement, crawl space or slab foundation available, please specify when ordering

    Price Code A

MATERIAL LIST Material List Available **SEE PAGE 81**

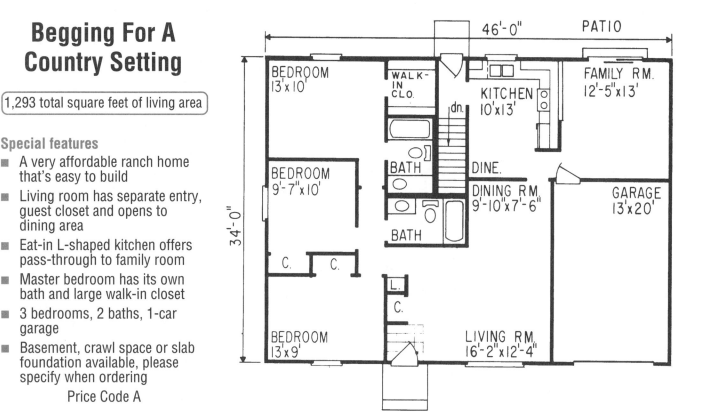

Plan #554-ES-147-1 basement
Plan #554-ES-147-2 crawl space & slab

Plan #554-ES-147-1 & 2

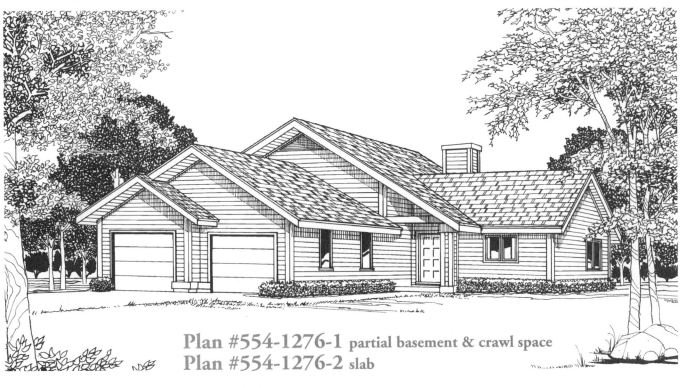

Plan #554-1276-1 partial basement & crawl space
Plan #554-1276-2 slab

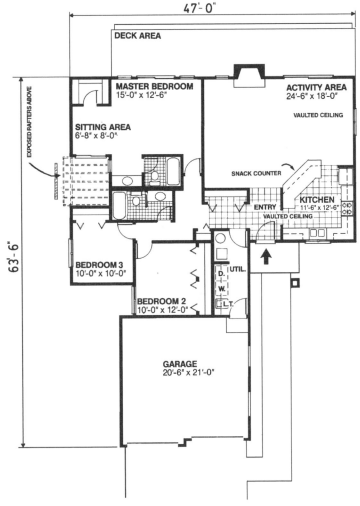

Floor plan labels:
- DECK AREA
- EXPOSED RAFTERS ABOVE
- MASTER BEDROOM 15'-0" x 12'-6"
- SITTING AREA 6'-8" x 8'-0"
- ACTIVITY AREA 24'-6" x 18'-0"
- VAULTED CEILING
- SNACK COUNTER
- ENTRY
- KITCHEN 11'-6" x 12'-6"
- VAULTED CEILING
- BEDROOM 3 10'-0" x 10'-0"
- UTIL.
- D.W. L.T.
- BEDROOM 2 10'-0" x 12'-0"
- GARAGE 20'-6" x 21'-0"
- 47'-0"
- 63'-6"

Multiple Gabled Roofs Add Drama

1,533 total square feet of living area

Special features

- Private deck outside the master bedroom sitting area
- Sloped ceilings add volume to the large activity area
- Activity room has fireplace, snack bar and shares access to the backyard with the master bedroom
- Convenient utility room located near the garage
- 3 bedrooms, 2 bath, 2-car garage
- Partial basement, crawl space or slab foundation available, please specify when ordering

Price Code B

Material List Available
SEE PAGE 81

WHITEFIELD

Plan #554-1104-1 basement
Plan #554-1104-2 crawl space & slab

Unique Sectioning For Entertaining

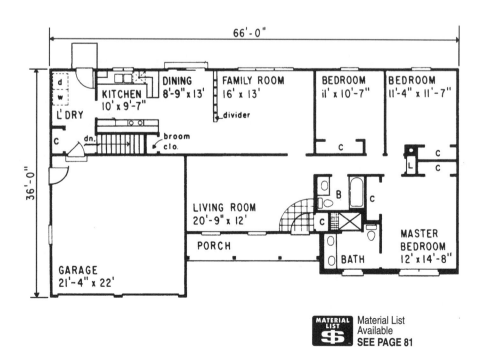

66'-0"

d	
w	
L'DRY	KITCHEN 10' x 9'-7"

DINING 8'-9" x 13'

FAMILY ROOM 16' x 13'
divider

BEDROOM il' x 10'-7"

BEDROOM 11'-4" x 11'-7"

c

36'-0"

dn. broom clo.

LIVING ROOM 20'-9" x 12'

B

c

c

L

c

MASTER BEDROOM 12' x14'-8"

PORCH

c

c

GARAGE 21'-4" x 22'

BATH

MATERIAL LIST Material List Available **SEE PAGE 81**

1,593 total square feet of living area

Special features

■ A welcoming porch invites you into a spacious living room

■ Kitchen and dining room opens to family room through wood balustrade

■ Master bedroom offers private bathroom and two closets

■ Laundry/mud room located directly off garage with convenient access to the backyard

■ 3 bedrooms, 2 baths, 2-car garage

■ Basement, crawl space or slab foundation available, please specify when ordering

Price Code B

Plan #554-1104-1 & 2

EVERGREEN

Great Room Forms Core Of This Home

2,076 total square feet of living area

Special features

- Vaulted great room fireplace flanked by windows and skylights that welcome the sun
- Kitchen leads to vaulted breakfast room and rear deck
- Study located off foyer provides great location for home office
- Large bay windows grace master bedroom and bath
- 3 bedrooms, 2 baths, 2-car garage
- Basement foundation

Price Code C

CUSTOMIZER KiT Customize This Plan SEE PAGE 4

MATERIAL LIST $ Material List Available SEE PAGE 81

J.N. HANSEN S.D.

Plan #554-0425

To order blueprints use the form on page 83 or call **1-800-DREAM HOME** *(373-2646)*

AVA

Rustic Stone Exterior

1,466 total square feet of living area

Special features

- Energy efficient home with 2" x 6" exterior walls
- Foyer separates the living room from the dining room and contains a generous coat closet
- Large living room with corner fireplace, bay window and pass-through to the kitchen
- Informal breakfast area opens out to large terrace through sliding glass doors which lets light into area
- Master bedroom has a large walk-in closet and private bath
- 3 bedrooms, 2 baths, 2-car garage
- Basement foundation, drawings also include slab foundation

Price Code A

56'-4"

49'-8"

Br 3
10-4x
10-0

Br 2
13-4x10-0

MBr
14-10x14-4

Kit
11-0x9-0

Brk
8-8x
9-0

Porch

Living
14-10x14-4

Dn

Dining
10-0x11-0

D
W

shelf

Porch depth 6-0

Garage
20-0x19-6

CUSTOMIZER KiT Customize This Plan SEE PAGE 4

MATERIAL LIST $ Material List Available SEE PAGE 81

Plan #554-0679

Wonderful Entertaining Possibilities

1,200 total square feet of living area

Special features

- Large living and dining rooms are completely open to one another with lots of space for large family gatherings
- The cozy kitchen has a half-wall open to dining room offering lots of entertaining possibilities
- All three bedrooms have nice sized closets and share a hall bath
- An optional second bath is available at rear of home
- 3 bedrooms, 1 bath, optional 2-car garage
- Basement, crawl space or slab foundation available, please specify when ordering

Price Code A

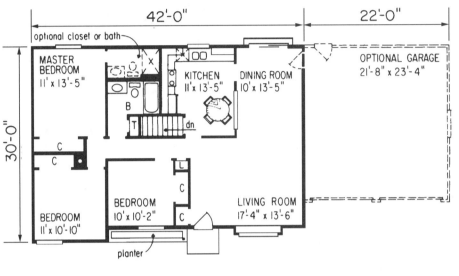

MATERIAL LIST $ Material List Available **SEE PAGE 81**

Plan #554-1199-1 basement
Plan #554-1199-2 crawl space & slab

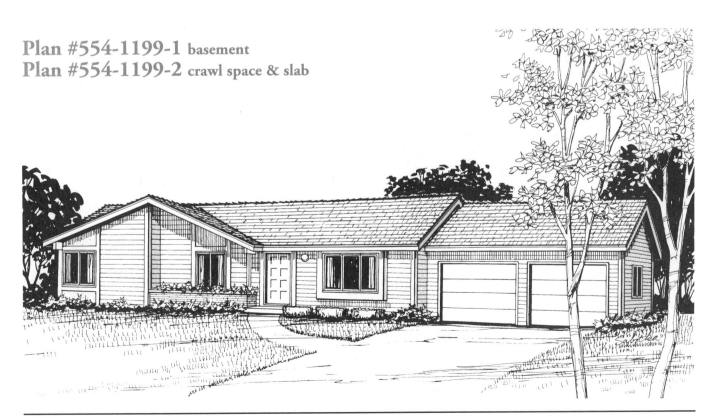

Plan #554-1199-1 & 2

COUNTRYSIDE

Affordable Upscale, Amenity Full

1,643 total square feet of living area

Special features

- Family room has vaulted ceiling, open staircase and arched windows allowing for plenty of light
- Kitchen captures full use of space, with pantry, storage, ample counter space and work island
- Large closets and storage areas throughout
- Roomy master bath has a skylight for natural lighting plus separate tub and shower
- Rear of house provides ideal location for future screened-in porch
- 3 bedrooms, 2 baths, 2-car side entry garage
- Basement foundation, drawings also include slab and crawl space foundations

Price Code B

CUSTOMIZER KIT Customize This Plan SEE PAGE 4

MATERIAL LIST $ Material List Available SEE PAGE 81

Plan #554-0172

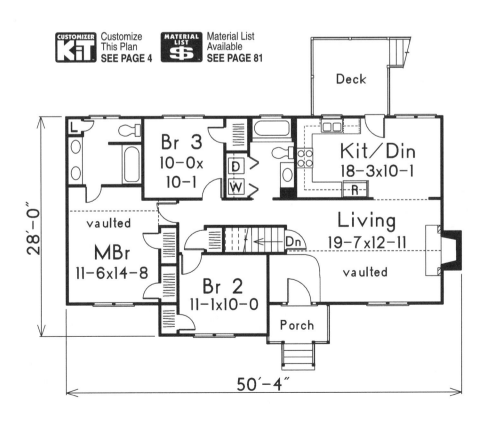

CUSTOMIZER KIT Customize This Plan **SEE PAGE 4**

MATERIAL LIST $ Material List Available **SEE PAGE 81**

Floor plan labels:

Deck

Br 3
10-0x
10-1

Kit/Din
18-3x10-1

D
W

R

Living
19-7x12-11

vaulted

MBr
11-6x14-8

Dn

Br 2
11-1x10-0

Porch

vaulted

28'-0"

50'-4"

Compact Home For Functional Living

1,220 total square feet of living area

Special features

- Vaulted ceilings add luxury to living room and master suite
- Spacious living room accented with a large fireplace and hearth
- Gracious dining area is adjacent to the convenient wrap-around kitchen
- Washer and dryer handy to the bedrooms
- Covered porch entry adds appeal
- Rear sun deck adjoins dining area
- 3 bedrooms, 2 baths, 2-car drive under garage
- Basement foundation

Price Code A

Plan #554-0173

BELLEVUE

Attractive And Spacious Brick Ranch

Plan #554-P-130-1 basement
Plan #554-P-130-2 crawl space & slab

1,778 total square feet of living area

Special features

- Formal entryway leads into large living room
- Family room with fireplace opens to patio
- Designed for practicality, kitchen/dining area are adjacent to the mud room/lavatory area that opens to the garage
- Designed to allow additional bedroom if future expansion is desired
- Future fourth bedroom has an additional 266 square feet of living area
- 3 bedrooms, 2 1/2 baths, 2-car garage
- Basement, crawl space or slab foundation available, please specify when ordering

Price Code B

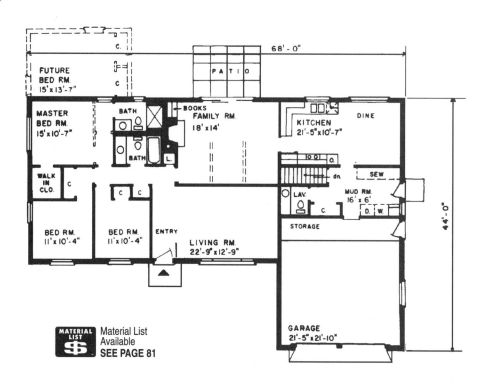

MATERIAL LIST $ Material List Available **SEE PAGE 81**

Plan #554-P-130-1 & 2

To order blueprints use the form on page 83 or call **1-800-DREAM HOME** (373-2646)

Smartly Elegant

1,907 total square feet of living area

Special features

- Activity area with fireplace opens to dining room
- Sun room off activity area leads to deck
- Laundry room conveniently located in bedroom wing of home
- Two bedrooms share a full bath
- Master bedroom suite features access to the sun room plus a deluxe master bath with clerestory window and large closets
- 3 bedrooms, 2 baths, 2-car garage
- Partial basement/crawl space foundation

Price Code C

MATERIAL LIST Material List Available **SEE PAGE 81**

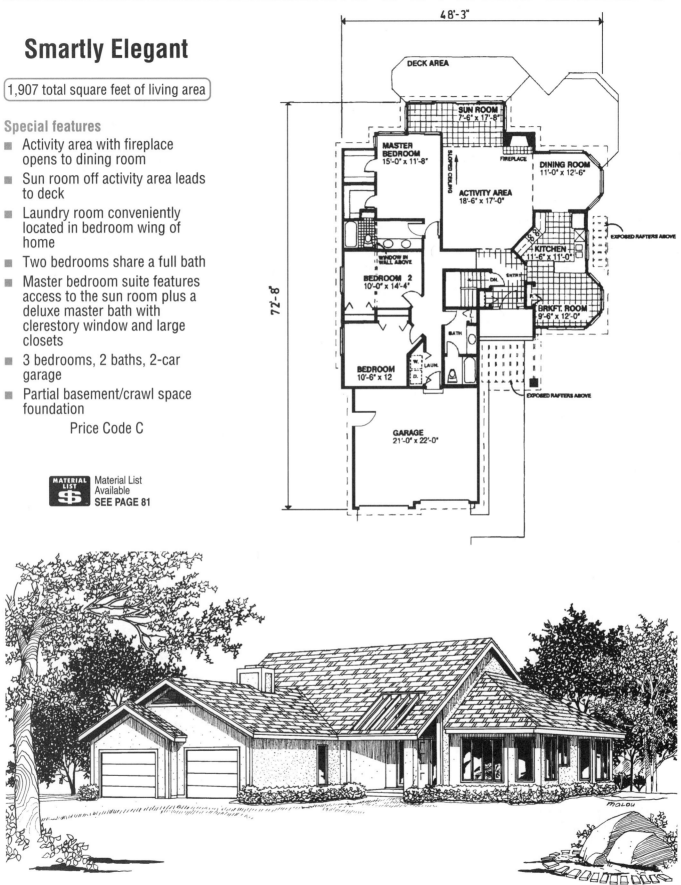

CEDARVILLE

Plan #554-1216-1 partial basement/crawl space
Plan #554-1216-2 crawl space & slab

Simply Country

1,668 total square feet of living area

Special features

- Simple but attractive styling ranch home is perfect for a narrow lot
- Front entry porch with entrance foyer with closet opens to living room
- Garage entrance to home leads to kitchen through mud room/laundry
- U-shaped kitchen opens to dining area and family room
- Three bedrooms are situated at the rear of the home with two full baths
- Master bedroom has walk-in closet
- 3 bedrooms, 2 baths, 2-car garage
- Partial basement/crawl space, crawl space or slab foundation available, please specify when ordering

Price Code B

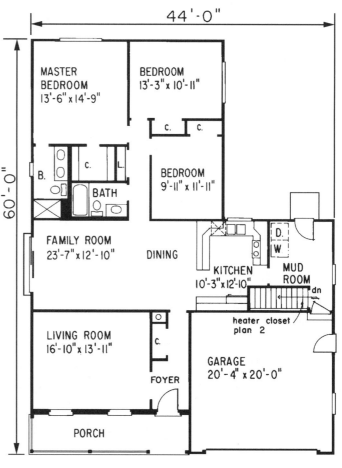

MATERIAL LIST $ Material List Available **SEE PAGE 81**

Plan #554-1216-1 & 2

WYNDHAM

Wonderful Great Room

Covered Porch

MBr
16-7x11-11
vaulted

plant shelf

Brk
10-5x8-11
vaulted

Great Rm
15-8x16-3
vaulted

Kit
7-9x
12-7

Br 2
13-3x9-11
vaulted

plant shelf

Dining
13-5x10-7
← Plant shelf

Garage
19-3x19-5

Br 3
13-3x11-4
vaulted

Br 4
10-11x
13-9
vaulted

Entry

66'-0"

45'-0"

R
P
D
W
L

1,865 total square feet of living area

Special features

- Large foyer opens into expansive dining/great room area
- Home features vaulted ceilings throughout
- Master suite features bath with double-bowl vanity, shower, tub and toilet in separate room for privacy
- 4 bedrooms, 2 baths, 2-car garage
- Slab foundation, drawings also include crawl space foundation

Price Code D

CUSTOMIZER KIT Customize This Plan SEE PAGE 4

MATERIAL LIST $ Material List Available SEE PAGE 81

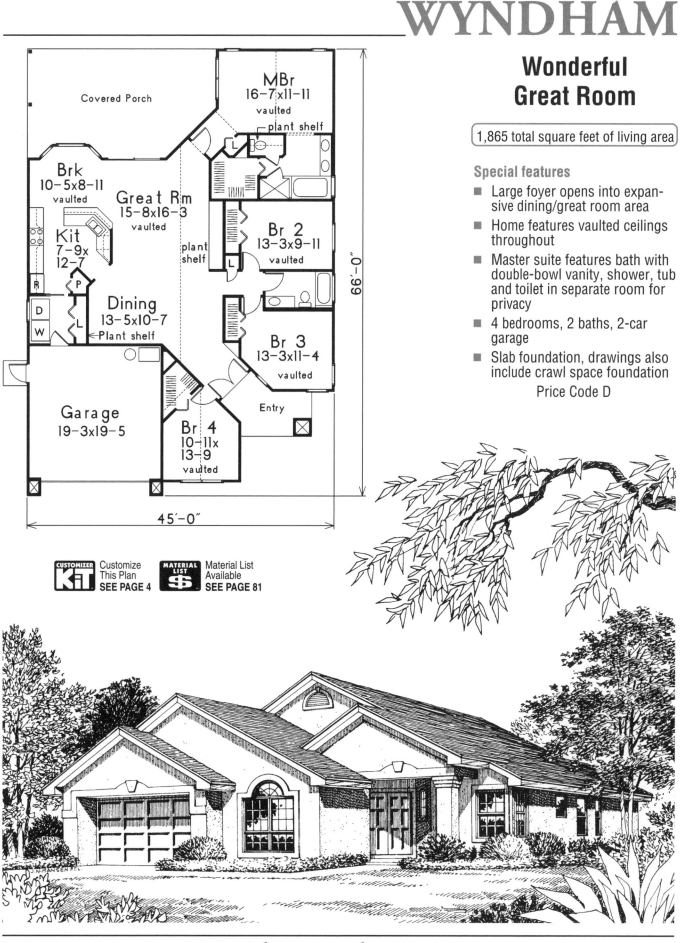

Plan #554-0335
To order blueprints use the form on page 83 or call **1-800-DREAM HOME** *(373-2646)*

VALLEYBROOK

Openness In A Split-Bedroom Ranch

1,574 total square feet of living area

Special features

- Foyer enters into open great room with corner fireplace and rear dining room with adjoining kitchen
- Two secondary bedrooms share a full bath
- Master bedroom has spacious private bath
- Garage accesses home through mud room/laundry
- 3 bedrooms, 2 baths, 2-car garage
- Basement foundation, drawings also include crawl space foundation

Price Code B

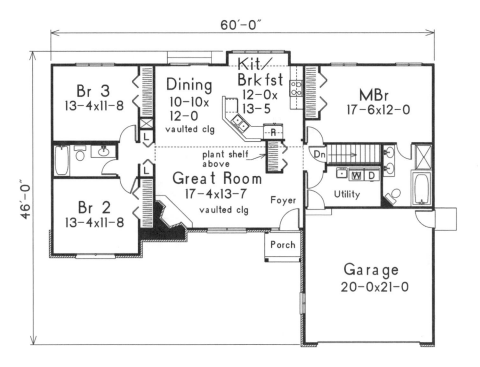

60'-0"

46'-0"

Br 3
13-4x11-8

Dining
10-10x
12-0
vaulted clg

Kit/
Brkfst
12-0x
13-5

MBr
17-6x12-0

Great Room
17-4x13-7
vaulted clg

plant shelf above

Dn

W D

Utility

Foyer

Br 2
13-4x11-8

Porch

Garage
20-0x21-0

Material List Available
SEE PAGE 81

Plan #554-1248

COURTLAND

Plan #554-1114-1 basement
Plan #554-1114-2 crawl space & slab

Impressive Ranch Features Attractive Courtyard

2,851 total square feet of living area

Special features

- Foyer with double-door entrance leads to unique sunken living room with patio view
- Multi-purpose room perfect for home office, hobby room or fifth bedroom
- Master bedroom boasts abundant closet space and access to patio
- Family room has access to kitchen and features a fireplace flanked by windows
- 4 bedrooms, 3 baths, 2-car garage
- Basement, crawl space or slab foundation available, please specify when ordering

Price Code E

Material List Available
SEE PAGE 81

Plan #554-1114-1&2

RICHELEY

Desirable Ranch With A Courtyard

1,691 total square feet of living area

Special features

- Cleverly located window in breakfast area adds charm to courtyard and view from kitchen
- Pass-through counter with bi-fold doors in kitchen provides convenience in serving to dining area
- Private bath and dual closets are featured in master bedroom
- 3 bedrooms, 2 baths, 2-car garage
- Basement, crawl space or slab foundation available, please specify when ordering

    Price Code B

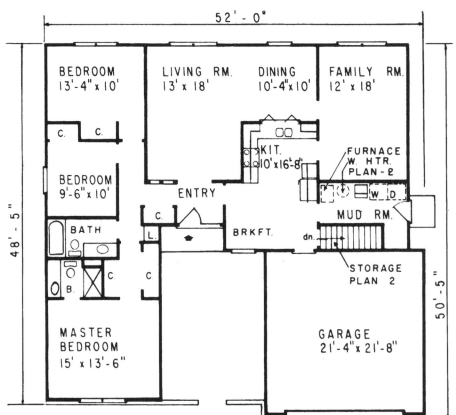

52' - 0"

48' - 5"

50' - 5"

BEDROOM 13'-4" x 10'

C.   C.

BEDROOM 9'-6" x 10'

BATH

B.

C

MASTER BEDROOM 15' x 13'-6"

LIVING RM. 13' x 18'

DINING 10'-4"x10'

FAMILY RM. 12' x 18'

KIT. 10'x16'-8"

FURNACE W. HTR. PLAN-2

W   D

ENTRY

C.

L.

C.

C.

BRKFT.

MUD RM.

dn.

STORAGE PLAN 2

GARAGE 21'-4" x 21'-8"

MATERIAL LIST $ Material List Available **SEE PAGE 81**

Plan #554-ES-115-1 basement
Plan #554-ES-115-2 crawl space & slab

Plan #554-1267-1 basement
Plan #554-1267-2 slab

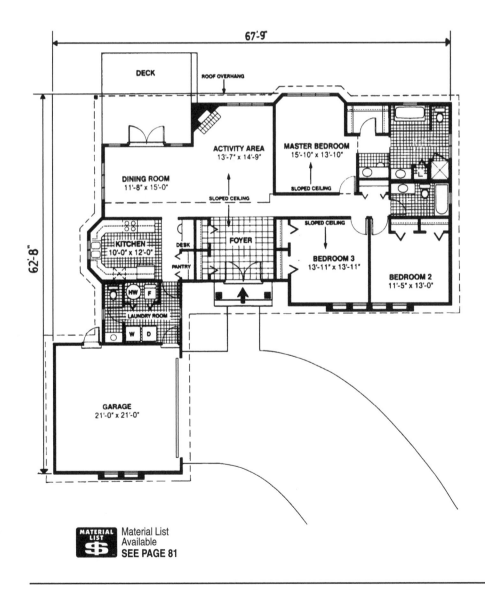

DECK

ROOF OVERHANG

67'-9"

ACTIVITY AREA
13'-7" x 14'-9"

MASTER BEDROOM
15'-10" x 13'-10"

DINING ROOM
11'-8" x 15'-0"

SLOPED CEILING

SLOPED CEILING

62'-8"

KITCHEN
10'-0" x 12'-0"

DESK

FOYER

SLOPED CEILING

PANTRY

BEDROOM 3
13'-11" x 13'-11"

BEDROOM 2
11'-5" x 13'-0"

HW F

LAUNDRY ROOM

W D

GARAGE
21'-0" x 21'-0"

MATERIAL LIST $ Material List Available **SEE PAGE 81**

Casual Ranch

1,800 total square feet of living area

Special features

- Comforts abound in this well-designed ranch
- Sunlit entryway leads to activity area with corner fireplace at rear of home
- U-shaped kitchen with built-in pantry and desk is adjacent to dining room with optional deck
- Large laundry area/powder room conveniently located adjacent to garage, just off kitchen
- Master bedroom features large walk-in closet, dressing area with make-up vanity and compartmented master bath with shower and raised tub
- Two additional bedrooms are served with a full bath
- 3 bedrooms, 2 1/2 baths, 2-car side entry garage
- Basement or slab foundation available, please specify when ordering

Price Code C

PARKSIDE

Exquisite Double-Door Entry

Plan #554-N301-1 basement/crawl space
Plan #554-N301-2 crawl space/slab

2,305 total square feet of living area

Special features

- Living room features large window for outdoor views
- Left side of home includes master suite with spacious walk-in closet and option of connecting to one of three remaining bedrooms for use as a den
- Right side of home includes family room with walk-out to patio adjacent to lovely dining room
- L-shaped counter in kitchen creates open atmosphere when connected with breakfast room
- 4 bedrooms 2 1/2 baths, 2-car side entry garage
- Basement/crawl space or crawl space/slab foundation available, please specify when ordering

Price Code D

MATERIAL LIST $ Material List Available **SEE PAGE 81**

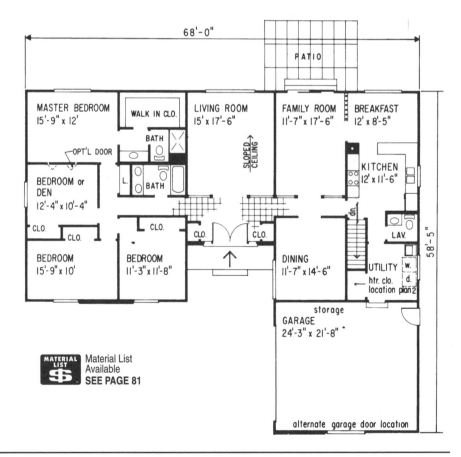

Plan #554-N301-1 & 2

Large Living Room With Dining Area

1,467 total square feet of living area

Special features

- Master bedroom has dressing area with vanity and private bath with toilet and tub
- Kitchen has direct access to garage and breakfast bar that overlooks the family room
- Family room has access to the outdoors and nearby laundry area
- Covered porch at rear of home is great for outdoor entertaining
- 3 bedrooms, 2 baths, 2-car garage
- Basement foundation

Price Code A

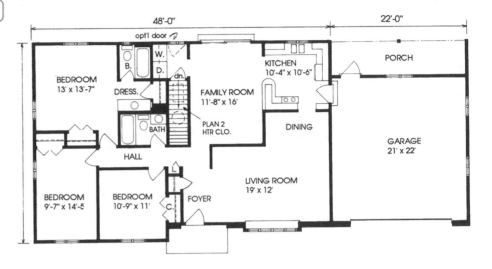

 Material List Available **SEE PAGE 81**

Plan #554-1291

*To order blueprints use the form on page 83 or call **1-800-DREAM HOME** (373-2646)*

KINSLEY

CUSTOMIZER KiT Customize This Plan **SEE PAGE 4**

MATERIAL LIST $S Material List Available **SEE PAGE 81**

Classic Exterior Employs Innovative Planning

1,791 total square feet of living area

Special features

- Vaulted great room and octagon-shaped dining area enjoy views of covered patio
- Kitchen features a pass-through to dining, center island, large walk-in pantry and breakfast room with large bay window
- Master bedroom is vaulted with sitting area
- 4 bedrooms, 2 baths, 2-car garage with storage
- Basement foundation

Price Code B

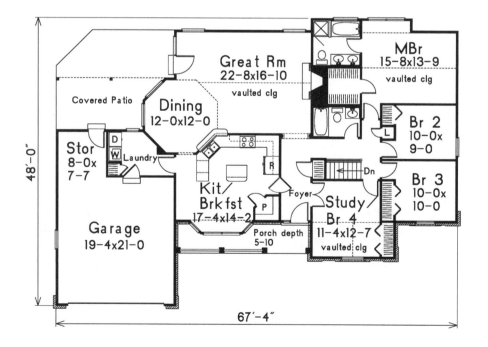

SANTA JENITA

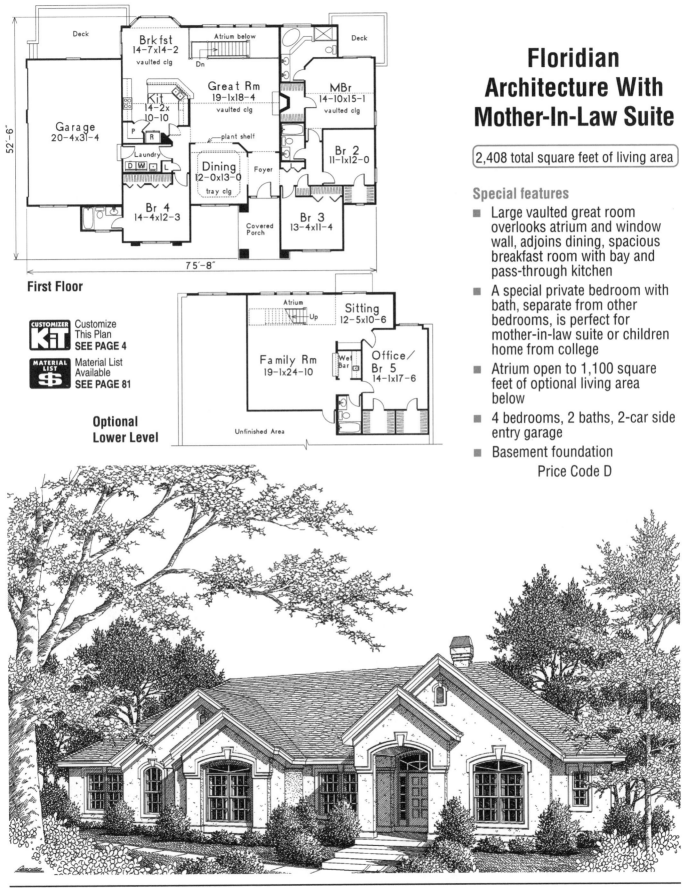

First Floor

Deck

Brkfst
14-7x14-2
vaulted clg

Atrium below

Dn

Deck

Kit
14-2x
10-10

Great Rm
19-1x18-4
vaulted clg

MBr
14-10x15-1
vaulted clg

Garage
20-4x31-4

P

R

plant shelf

Br 2
11-1x12-0

Laundry

D W L

Dining
12-0x13-0
tray clg

Foyer

Br 4
14-4x12-3

Covered
Porch

Br 3
13-4x11-4

52'-6"

75'-8"

CUSTOMIZER KiT Customize This Plan **SEE PAGE 4**

MATERIAL LIST $ Material List Available **SEE PAGE 81**

Optional Lower Level

Atrium

Up

Sitting
12-5x10-6

Family Rm
19-1x24-10

Wet Bar

Office/
Br 5
14-1x17-6

Unfinished Area

Floridian Architecture With Mother-In-Law Suite

2,408 total square feet of living area

Special features

- Large vaulted great room overlooks atrium and window wall, adjoins dining, spacious breakfast room with bay and pass-through kitchen

- A special private bedroom with bath, separate from other bedrooms, is perfect for mother-in-law suite or children home from college

- Atrium open to 1,100 square feet of optional living area below

- 4 bedrooms, 2 baths, 2-car side entry garage

- Basement foundation

    Price Code D

Plan #554-0730

HILLTOP

Palladian Windows Dominate Facade

1,500 total square feet of living area

Special features

- Living room features a cathedral ceiling and opens to breakfast room
- Breakfast room has a spectacular bay window and adjoins a well-appointed kitchen with generous storage
- Laundry is convenient to kitchen and includes a large closet
- Large walk-in closet gives the master bedroom abundant storage
- 3 bedrooms, 2 baths, 2-car garage
- Basement foundation

    Price Code B

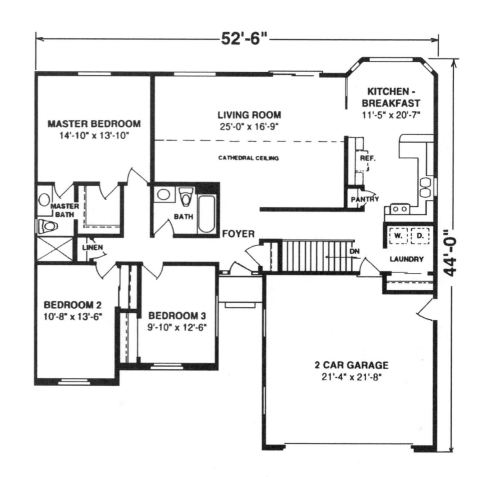

52'-6"

44'-0"

MASTER BEDROOM
14'-10" x 13'-10"

LIVING ROOM
25'-0" x 16'-9"

CATHEDRAL CEILING

KITCHEN - BREAKFAST
11'-5" x 20'-7"

REF.

PANTRY

MASTER BATH

BATH

LINEN

FOYER

DN

W. D.

LAUNDRY

BEDROOM 2
10'-8" x 13'-6"

BEDROOM 3
9'-10" x 12'-6"

2 CAR GARAGE
21'-4" x 21'-8"

Plan #554-1429

To order blueprints use the form on page 83 or call **1-800-DREAM HOME** (373-2646)

SUNRIDGE

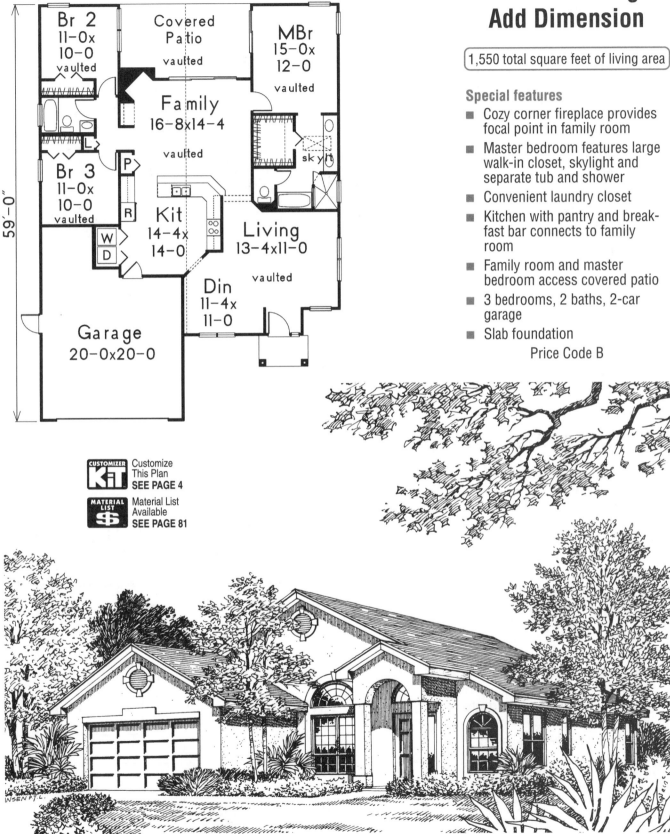

43'-0"

Br 2
11-0x
10-0
vaulted

Covered
Patio
vaulted

MBr
15-0x
12-0
vaulted

Family
16-8x14-4
vaulted

skylt

Br 3
11-0x
10-0
vaulted

P

R

Kit
14-4x
14-0

Living
13-4x11-0
vaulted

W
D

Din
11-4x
11-0

59'-0"

Garage
20-0x20-0

Vaulted Ceilings
Add Dimension

1,550 total square feet of living area

Special features

- Cozy corner fireplace provides focal point in family room
- Master bedroom features large walk-in closet, skylight and separate tub and shower
- Convenient laundry closet
- Kitchen with pantry and break- fast bar connects to family room
- Family room and master bedroom access covered patio
- 3 bedrooms, 2 baths, 2-car garage
- Slab foundation

Price Code B

CUSTOMIZER KiT Customize This Plan **SEE PAGE 4**

MATERIAL LIST $ Material List Available **SEE PAGE 81**

Plan #554-0357
To order blueprints use the form on page 83 or call **1-800-DREAM HOME** *(373-2646)*

59

CURTLAND

Appealing Ranch Has Attractive Front Dormers

1,642 total square feet of living area

Special features

- Walk-through kitchen boasts vaulted ceiling and corner sink overlooking family room
- Vaulted family room features cozy fireplace and access to rear patio
- Master bedroom includes sloped ceiling, walk-in closet and private bath
- 3 bedrooms, 2 baths, 2-car garage
- Basement foundation, drawings also include slab and crawl space foundations

    Price Code B

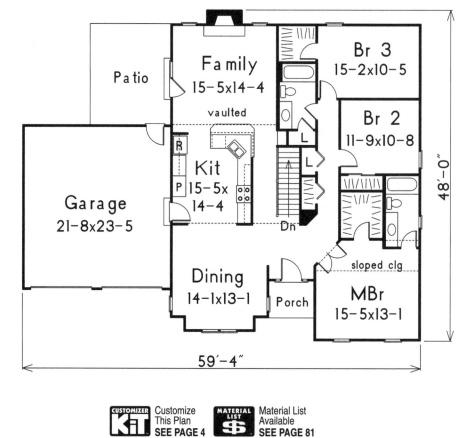

CUSTOMIZER KiT Customize This Plan SEE PAGE 4

MATERIAL LIST $ Material List Available SEE PAGE 81

Plan #554-0282

Sculptured Roof Line And Facade Add Charm

1,674 total square feet of living area

Special features

- Great room, dining area and kitchen, surrounded with vaulted ceiling, central fireplace and log bin
- Convenient laundry/mud room located between garage and family area with handy stairs to basement
- Easily expandable screened porch and adjacent patio with access from dining area
- Master bedroom features full bath with tub, separate shower and walk-in closet
- 3 bedrooms, 2 baths, 2-car garage
- Basement foundation, drawings also include crawl space and slab foundations

Price Code B

Customizer Kit — Customize This Plan SEE PAGE 4

Material List Available SEE PAGE 81

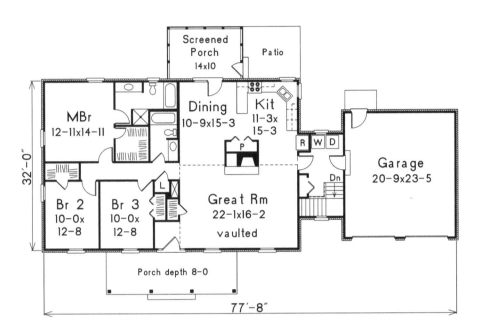

Screened Porch 14x10

Patio

MBr 12-11x14-11

Dining 10-9x15-3

Kit 11-3x 15-3

R W D

Garage 20-9x23-5

Dn

Br 2 10-0x 12-8

Br 3 10-0x 12-8

Great Rm 22-1x16-2 vaulted

32'-0"

Porch depth 8-0

77'-8"

VILLA SUPREME

Vaulted Villa

Plan #554-1263-1 partial basement/crawl space
Plan #554-1263-2 slab

2,155 total square feet of living area

Special features

- Unique roof angles make this an eye-catching design
- Foyer entrance angles to access living room, activity area, and dining room
- Vaulted ceilings featured in the living room and activity room
- Activity room also includes fireplace
- Kitchen with center island opens to breakfast bay area and dining room
- Master bedroom with bay window includes a raised tub master bath and two walk-in closets
- Two additional bedrooms share full bath
- 3 bedrooms, 2 baths, 2-car garage
- Partial basement/crawl space or slab foundation available, please specify when ordering

Price Code C

MATERIAL LIST Material List Available **SEE PAGE 81**

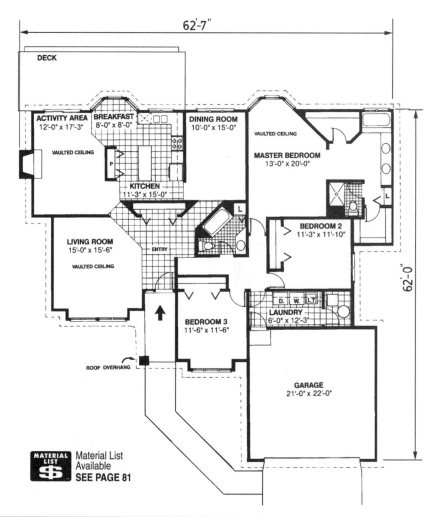

Plan #554-1263-1 & 2

FLORENCE

Comfortable One-Story Country Home

1,367 total square feet of living area

Special features

■ Neat front porch shelters the entrance

■ Dining room has full wall of windows and convenient storage area

■ Breakfast area leads to the rear terrace through sliding doors

■ Large living room with high ceiling, skylight and fireplace

■ 3 bedrooms, 2 baths, 2-car garage

■ Basement foundation, drawings also include slab foundation

Price Code A

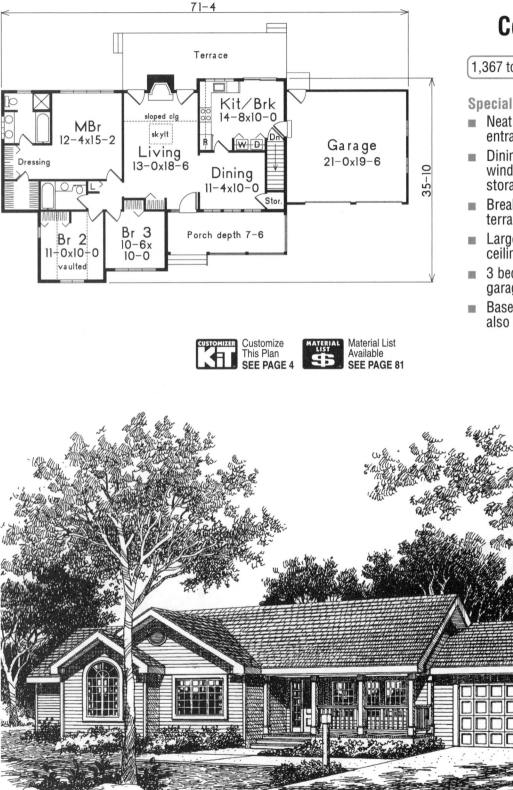

71-4

Terrace

MBr
12-4x15-2

sloped clg

skylt

Living
13-0x18-6

Kit/Brk
14-8x10-0

Dressing

Dining
11-4x10-0

W D

Dn

Stor.

Garage
21-0x19-6

35-10

Br 2
11-0x10-0
vaulted

Br 3
10-6x
10-0

Porch depth 7-6

CUSTOMIZER KiT Customize This Plan **SEE PAGE 4**

MATERIAL LIST $ Material List Available **SEE PAGE 81**

Plan #554-0676

ROYALOAK

Ideal For A Shallow Lot

1,414 total square feet of living area

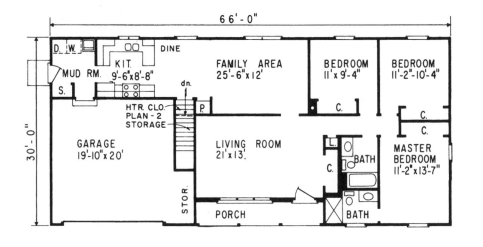

Special features

- Charming U-shaped kitchen offers lots of space with adjacent dining area
- Convenient to kitchen is a spacious laundry room and stair to basement
- All bedrooms enjoy ample closet storage
- Spacious living room with ample closet space
- 3 bedrooms, 2 baths, 2-car garage
- Basement, crawl space or slab foundation available, please specify when ordering

Price Code A

Material List Available
SEE PAGE 81

Plan #554-ES-112-1 basement
Plan #554-ES-112-2 crawl space & slab

Plan #554-ES-112-1 & 2

SUGARCREEK

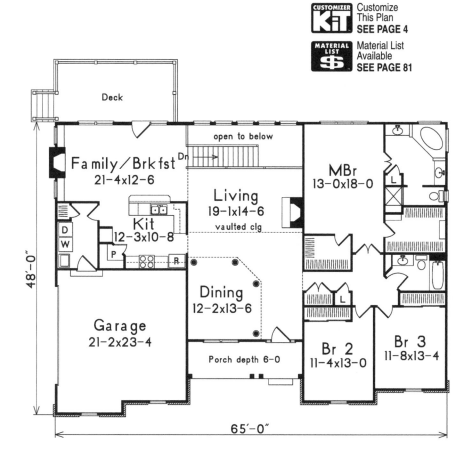

CUSTOMIZER KiT Customize This Plan **SEE PAGE 4**

MATERIAL LIST $ Material List Available **SEE PAGE 81**

Bright, Spacious Plan With Many Features

2,308 total square feet of living area

Special features

- Efficient kitchen designed with many cabinets and large walk-in pantry adjoins family/breakfast area featuring beautiful fireplace

- Dining area has architectural colonnades that separate it from living area while maintaining spaciousness

- Enter master suite through double-doors and find double walk-in closets and beautiful luxurious bath

- Living room includes vaulted ceiling, fireplace and a sunny atrium window wall creating a dramatic atmosphere

- 3 bedrooms, 2 baths, 2-car side entry garage

- Walk-out basement foundation

Price Code D

BRADBURY

Gabled Facade For A Lasting Impression

| 2,086 total square feet of living area |

Special features

- An angled foyer leads to vaulted living room with sunken floor

- Dining room, activity room, nook and kitchen all have vaulted ceilings

- Skillfully designed kitchen features an angled island with breakfast bar

- Master bedroom is state-of-the-art with luxury bath, giant walk-in closet and deck area for hot tub

- 3 bedrooms, 2 baths, 2-car garage

- Partial basement/crawl space foundation

Price Code C

MATERIAL LIST $ Material List Available **SEE PAGE 81**

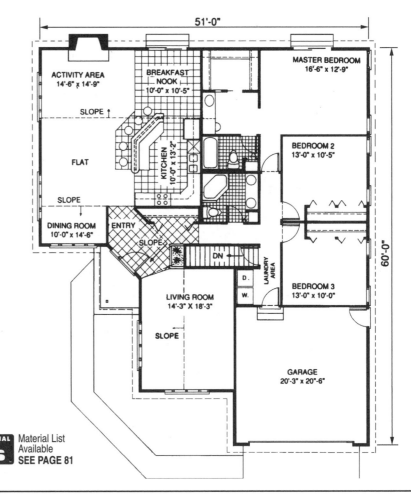

51'-0"

ACTIVITY AREA
14'-6" x 14'-9"

BREAKFAST NOOK
10'-0" x 10'-5"

MASTER BEDROOM
16'-6" x 12'-9"

SLOPE ↑

FLAT

KITCHEN
10'-0" x 13'-2"

BEDROOM 2
13'-0" x 10'-5"

SLOPE ↓

DINING ROOM
10'-0" x 14'-6"

ENTRY

SLOPE

DN

LAUNDRY AREA

BEDROOM 3
13'-0" x 10'-0"

LIVING ROOM
14'-3" X 18'-3"

D.

W.

SLOPE

GARAGE
20'-3" x 20"-6"

60'-0"

Plan #554-1266

ROSEBURY

Plan #554-1400-1 basement
Plan #554-1400-2 crawl space

Basic Serenity

1,102 total square feet of living area

Special features

- Compact design with a dressy exterior
- Spacious living room with separate entry and coat closet
- Eat-in kitchen with first floor laundry
- Master bedroom and second bedroom share roomy full bath
- 2 bedrooms, 1 bath, 2-car garage
- Basement or crawl space foundation available, please specify when ordering

Price Code AA

 Material List Available **SEE PAGE 81**

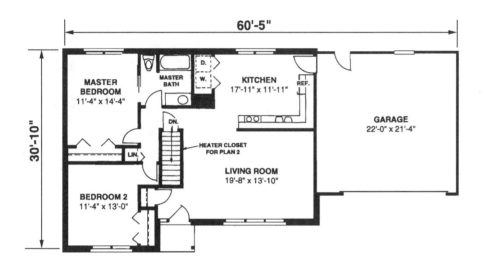

DAVIDSON

Lighted Charm

1,540 total square feet of living area

Special features

- Porch entrance into foyer leads to an impressive dining area with full window and a half-circle window above

- Kitchen/breakfast room features a center island and cathedral ceiling

- Great room with cathedral ceiling and exposed beams accessible from foyer

- Master bedroom includes full bath and walk-in closet

- Two additional bedrooms share a full bath

- 3 bedrooms, 2 baths, 2-car garage

- Basement foundation, drawings also include crawl space and slab foundations

    Price Code B

 Material List Available **SEE PAGE 81**

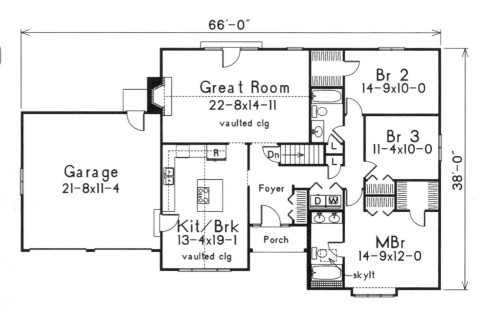

66'-0"

38'-0"

Great Room
22-8x14-11
vaulted clg

Br 2
14-9x10-0

Br 3
11-4x10-0

Garage
21-8x11-4

Kit/Brk
13-4x19-1
vaulted clg

Foyer

Porch

MBr
14-9x12-0

sky lt

Dn

Plan #554-1220

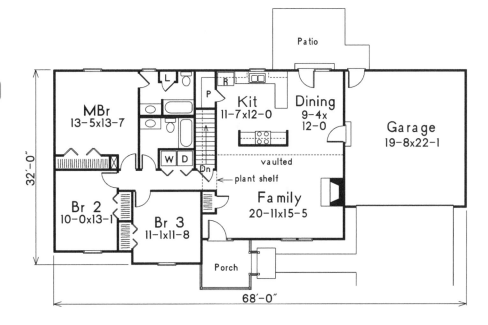

WYDOWN

Spacious And Open Family Living Area

1,416 total square feet of living area

Special features

- Family room includes fireplace, elevated plant shelf and vaulted ceiling
- Patio is accessible from dining area and garage
- Centrally located laundry area
- Oversized walk-in pantry
- 3 bedrooms, 2 baths, 2-car garage
- Basement foundation, drawings also include crawl space and slab foundations

Price Code A

CUSTOMIZER KiT Customize This Plan **SEE PAGE 4**

MATERIAL LIST $ Material List Available **SEE PAGE 81**

DONNELLY

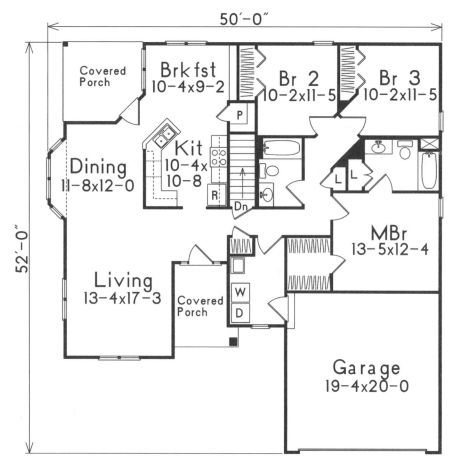

50'-0"

52'-0"

Covered Porch

Brkfst
10-4x9-2

Br 2
10-2x11-5

Br 3
10-2x11-5

Kit
10-4x
10-8

Dining
11-8x12-0

MBr
13-5x12-4

Living
13-4x17-3

Covered Porch

Garage
19-4x20-0

Designed For Handicap Access

1,578 total square feet of living area

Special features

- Plenty of closet, linen and storage space
- Covered porches in the front and rear of home add charm to this design
- Open floor plan with unique angled layout
- 3 bedrooms, 2 baths, 2-car garage
- Basement foundation

   Price Code B

CUSTOMIZER KIT Customize This Plan **SEE PAGE 4**

MATERIAL LIST $ Material List Available **SEE PAGE 81**

Plan #554-0741

OAKRIDGE

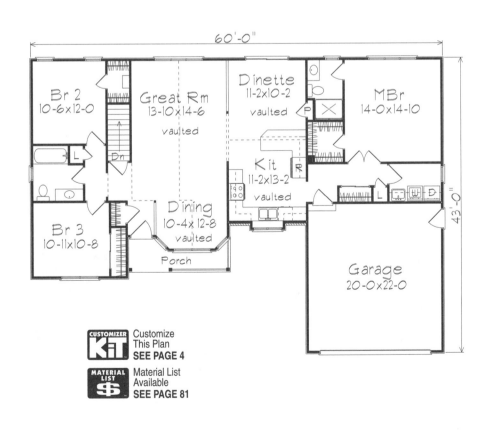

60'-0"

Br 2
10-6x12-0

Great Rm
13-10x14-6
vaulted

Dn

Dinette
11-2x10-2
vaulted

MBr
14-0x14-10

P

R

Kit
11-2x13-2
vaulted

Dining
10-4x12-8
vaulted

Porch

Br 3
10-11x10-8

43'-0"

L

W D

Garage
20-0x22-0

CUSTOMIZER KiT Customize This Plan **SEE PAGE 4**

MATERIAL LIST $ Material List Available **SEE PAGE 81**

Central Living Area Keeps Bedrooms Private

1,546 total square feet of living area

Special features

- Spacious, open rooms create casual atmosphere
- Master suite secluded for privacy
- Dining room features large bay window
- Kitchen/dinette combination offers access to the outdoors
- Large laundry room includes convenient sink
- 3 bedrooms, 2 baths, 2-car garage
- Basement foundation

  Price Code B

Plan #554-0382

BRIARFIELD

An Enhancement To Any Neighborhood

> 1,440 total square feet of living area

Special features

- Foyer adjoins massive-sized great room with sloping ceilings and tall masonry fireplace

- Kitchen adjoins spacious dining room and features pass-through breakfast bar

- Master suite enjoys private bath and two closets

- An oversized two-car side entry garage offers plenty of storage for bicycles, lawn equipment, etc.

- 3 bedrooms, 2 baths, 2-car side entry garage

- Basement, crawl space or slab foundation available, please specify when ordering

Price Code A

Material List Available
SEE PAGE 81

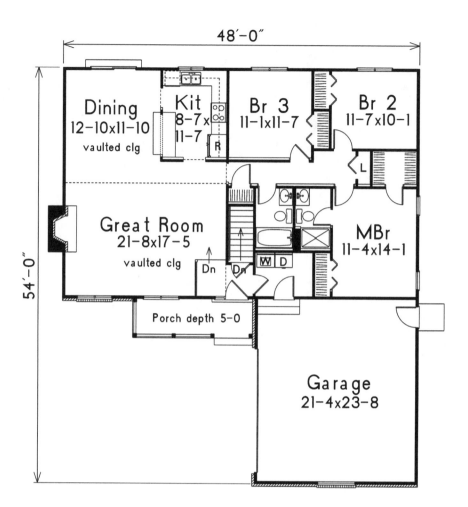

48'-0"

54'-0"

Dining
12-10x11-10
vaulted clg

Kit
8-7x
11-7

Br 3
11-1x11-7

Br 2
11-7x10-1

Great Room
21-8x17-5
vaulted clg

MBr
11-4x14-1

Dn

Dn

W D

Porch depth 5-0

Garage
21-4x23-8

FONTANA

Plan #554-1274-1 partial basement/crawl space
Plan #554-1274-2 partial slab/crawl space

Contemporary Excellence

2,180 total square feet of living area

Special features

- Large impressive entry for receiving guests
- Activity and dining rooms with vaulted ceiling, fireplace, wet bar and expansive bay windows are second to none
- Master bedroom and bath have been designed on a grand scale
- All bedrooms feature vaulted ceilings and spacious closets
- 3 bedrooms, 2 baths, 2-car garage
- Partial basement/crawl space or partial slab/crawl space foundation available, please specify when ordering

Price Code C

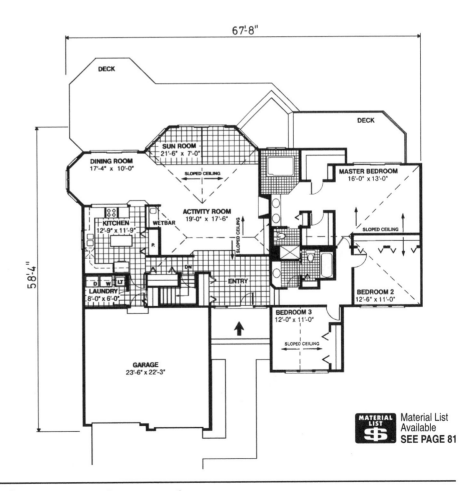

MATERIAL LIST $ — Material List Available **SEE PAGE 81**

Plan #554-1274-1 & 2

To order blueprints use the form on page 83 or call **1-800-DREAM HOME** (373-2646)

73

CONWAY

Unique Contemporary For The Handicapped

1,340 total square feet of living area

Special features

- Striking contemporary design offers complete accessibility for the handicapped
- Deep two-car garage provides ramp access into utility room
- Front entrance is via ramped front porch into foyer
- Combined kitchen/dining/living area illuminated by three large roof windows
- Rear of home features two separated covered decks for the ultimate in relaxation and outdoor leisure
- 2 bedrooms, 2 baths, 2-car garage
- Slab foundation

Price Code A

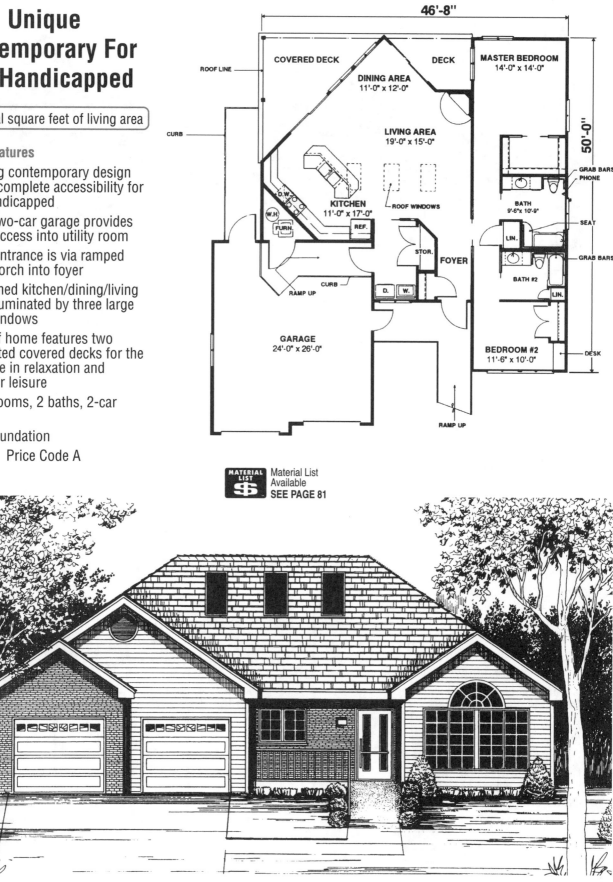

MATERIAL LIST Material List Available SEE PAGE 81

COUNTRY MANOR

Excellent Ranch For Country Setting

2,758 total square feet of living area

Special features

- Vaulted great room excels with fireplace, wet bar, plant shelves and skylights
- Fabulous master suite enjoys a fireplace, large bath, walk-in closet and vaulted ceiling
- Trendsetting kitchen/breakfast room adjoins spacious screened porch
- Convenient office near kitchen is perfect for computer room, hobby enthusiast or fifth bedroom
- 4 bedrooms, 2 1/2 baths, 3-car side entry garage
- Basement foundation

    Price Code E

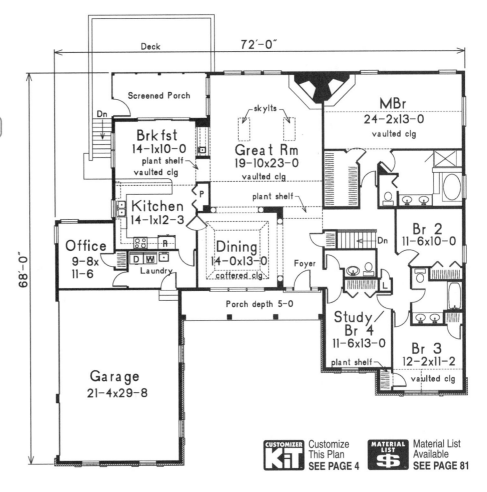

CUSTOMIZER **KiT** Customize This Plan **SEE PAGE 4**

MATERIAL **LIST** **$** Material List Available **SEE PAGE 81**

Plan #554-0705

SYCAMORE

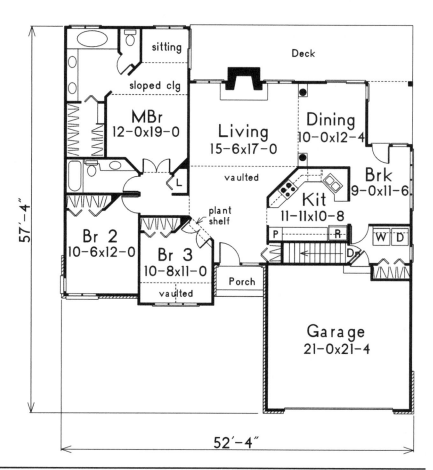

CUSTOMIZER KiT Customize This Plan **SEE PAGE 4**

MATERIAL LIST $ Material List Available **SEE PAGE 81**

Open Ranch Design Gives Expansive Look

1,630 total square feet of living area

Special features

- Crisp facade and full windows front and back offer open viewing

- Wrap-around rear deck is accessible from breakfast room, dining room and master bedroom

- Vaulted ceiling in living room and master bedroom

- Sitting area and large walk-in closet complement master bath

- Master bedroom has a private sitting area

- 3 bedrooms, 2 baths, 2-car garage

- Basement foundation

    Price Code B

Plan #554-0161

High Ceilings Create A Feeling Of Luxury

1,707 total square feet of living area

Special features

- The formal living room off the entry hall has a high sloping ceiling and prominent fireplace
- Kitchen and breakfast areas allow access to garage and rear porch
- Garage with oversized storage/ work area provides direct access to the kitchen
- Master bedroom has impressive vaulted ceiling, luxurious master bath, large walk-in closet and separate tub and shower
- Utility room conveniently located near bedrooms
- 3 bedrooms, 2 baths, 2-car garage
- Slab foundation

  Price Code C

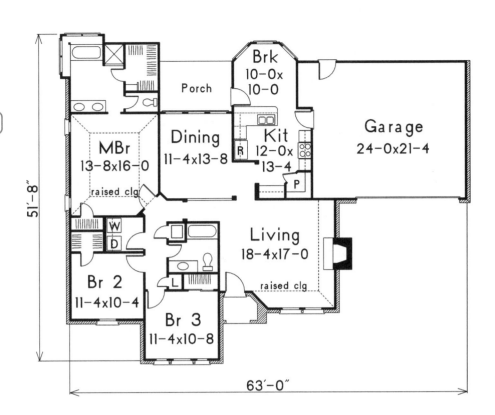

CUSTOMIZER KIT Customize This Plan **SEE PAGE 4**

MATERIAL LIST $ Material List Available **SEE PAGE 81**

Plan #554-0212

BAXTER

Ideal Ranch For A Narrow Lot

1,624 total square feet of living area

Special features

- Complete master bedroom suite with private entry from the outdoors

- Garage adjacent to utility room with convenient storage closet

- Large family/dining area with fireplace and porch access

- Pass-through kitchen opens directly to cozy breakfast area

- 3 bedrooms, 2 baths, 2-car side entry garage

- Basement foundation, drawings also include crawl space and slab foundations

Price Code B

Customize This Plan SEE PAGE 4

Material List Available SEE PAGE 81

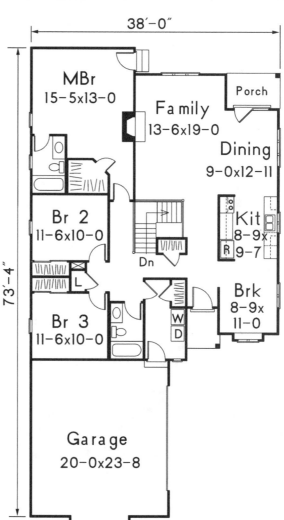

38'-0"

73'-4"

MBr
15-5x13-0

Family
13-6x19-0

Porch

Dining
9-0x12-11

Br 2
11-6x10-0

Kit
8-9x
9-7

L

Dn

Brk
8-9x
11-0

Br 3
11-6x10-0

W
D

Garage
20-0x23-8

Plan #554-0281

LINDENWOOD

Comfortable Family Living In This Ranch

1,994 total square feet of living area

Special features

- Convenient entrance from the garage into the main living space through the utility room
- Standard 9' ceilings, bedroom #2 features a 12' vaulted ceiling and a 10' ceiling in the dining room
- Master bedroom offers a full bath with oversized tub, separate shower and walk-in closet
- Entry leads to formal dining room and attractive living room with double French doors and fireplace
- 3 bedrooms, 2 baths, 2-car garage
- Slab foundation

Price Code D

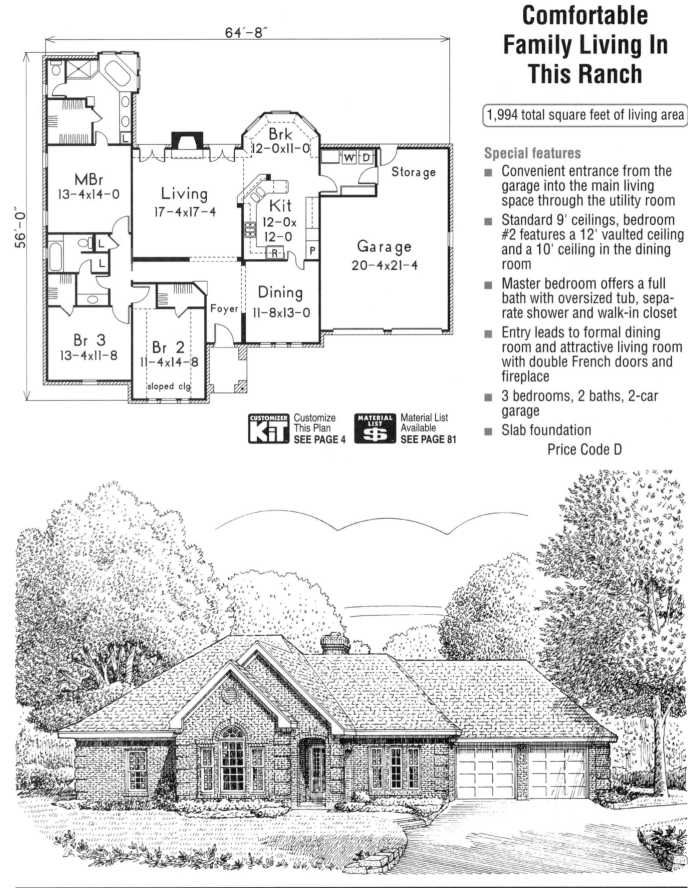

CUSTOMIZER KIT Customize This Plan SEE PAGE 4

MATERIAL LIST $ Material List Available SEE PAGE 81

64'-8"

56'-0"

MBr 13-4x14-0

Living 17-4x17-4

Brk 12-0x11-0

W D

Storage

Kit 12-0x 12-0

Garage 20-4x21-4

Dining 11-8x13-0

Foyer

Br 3 13-4x11-8

Br 2 11-4x14-8

sloped clg

Plan #554-0244

Our Blueprint Packages Offer...

Quality plans for building your future, with extras that provide unsurpassed value, ensure good construction and long-term enjoyment.

A quality home - one that looks good, functions well, and provides years of enjoyment - is a product of many things - design, materials, craftsmanship. But it's also the result of out-standing blueprints - the actual plans and specifications that tell the builder exactly how to build your home.

And with our BLUEPRINT PACKAGES you get the absolute best. A complete set of blueprints is available for every design in this book. These "working drawings," are highly detailed, resulting in two key benefits:

- *Better understanding by the contractor of how to build your home, and...*
- *More accurate construction estimates.*

When you purchase one of our designs, you'll receive all of the BLUEPRINT components shown here - elevations, foundation plan, floor plans, cross-sections, and details. Other helpful building aids are also available to help make your dream home a reality.

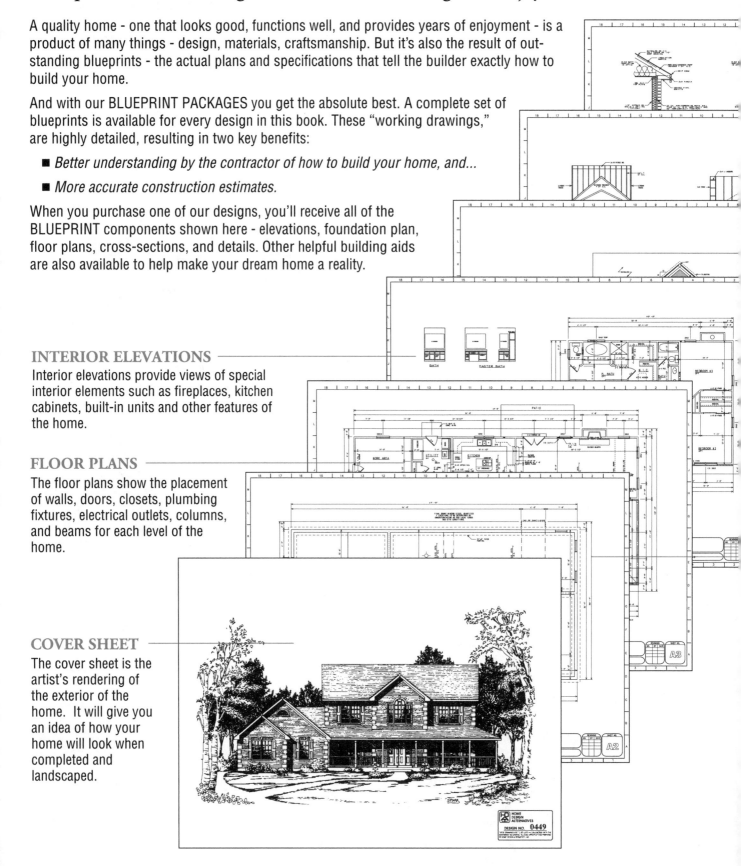

INTERIOR ELEVATIONS

Interior elevations provide views of special interior elements such as fireplaces, kitchen cabinets, built-in units and other features of the home.

FLOOR PLANS

The floor plans show the placement of walls, doors, closets, plumbing fixtures, electrical outlets, columns, and beams for each level of the home.

COVER SHEET

The cover sheet is the artist's rendering of the exterior of the home. It will give you an idea of how your home will look when completed and landscaped.

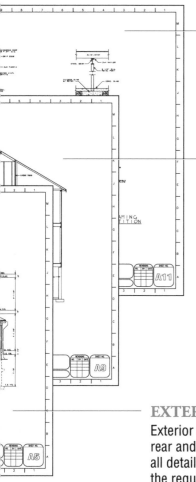

DETAILS

Details show how to construct certain components of your home, such as the roof system, stairs, deck, etc.

CROSS SECTIONS

Cross sections show detail views of the home as if it were sliced from the roof to the foundation. This sheet shows important areas such as load-bearing walls, stairs, joists, trusses and other structural elements, which are critical for proper construction.

EXTERIOR ELEVATIONS

Exterior elevations illustrate the front, rear and both sides of the house, with all details of exterior materials and the required dimensions.

FOUNDATION PLAN

The foundation plan shows the layout of the basement, crawl space, slab, or pier foundation. All necessary notations and dimensions are included. See plan page for the foundation types included. If the home plan you choose does not have your desired foundation type, our Customer Service Representatives can advise you on how to customize your foundation to suit your specific needs or site conditions.

GENERAL BUILDING SPECIFICATIONS

This document outlines the technical requirements for proper construction such as the strength of materials, insulation ratings, allowable loading conditions, etc.

Other Helpful Building Aids...

Your Blueprint Package will contain all the necessary construction information to build your home. We also offer the following products and services to save you time and money in the building process.

Material List

Material lists are available for many of our plans. Each list gives you the quantity, dimensions and description of the building materials necessary to construct your home. You'll get faster and more accurate bids from your contractor and material suppliers, and you'll save money by paying for only the materials you need. Refer to the order form on page 83 for pricing.

Customizer Kit ™

Many of the designs in this book can be customized using our exclusive Customizer Kit. It's your guide to custom designing your home. It leads you through all the essential design decisions and provides the necessary tools for you to clearly show the changes you want made. Customizer Kits are available for most designs in this book. For more information about this exclusive product see page 4.

Detail Plan Packages:
FRAMING, PLUMBING AND ELECTRICAL PLAN PACKAGES

Three separate packages offer homebuilders details for constructing various foundations; numerous floor, wall and roof framing techniques; simple to complex residential wiring; sump and water softener hookups; plumbing connection methods; installation of septic systems and more. Each package includes three-dimensional illustrations and a glossary of terms. Purchase one or all three. Refer to the order form on page 83 for pricing.

The Legal Kit ™

Our Legal Kit provides contracts and legal forms to help protect you from the potential pitfalls inherent in the building process. The Kit supplies commonly used forms and contracts suitable for homeowners and builders. It can save you a considerable amount of time and help protect you and your assets during and after construction. Refer to the order form on page 83 for pricing.

Rush Delivery

Most orders are processed within 24 hours of receipt. Please allow 7 working days for delivery. If you need to place a rush order, please call us by 11:00 a.m. CST and ask for overnight or second day service.

Technical Assistance

If you have questions, call our technical support line at 1-314-770-2228 between 8:00 a.m. and 5:00 p.m. CST. Whether it involves design modifications or field assistance, our designers are extremely familiar with all of our designs and will be happy to help you. We want your home to be everything you expect it to be.

HOME DESIGN ALTERNATIVES, INC.

What Kind Of Plan Package Do You Need?

Now that you've found the home plan you've been looking for, here are some suggestions on how to make your Dream Home a reality. To get started, order the type of plans that fit your particular situation.

YOUR CHOICES:

The One-set package - This single set of blueprints is offered so you can study or review a home in greater detail. But a single set is never enough for construction and it's a copyright violation to reproduce blueprints.

The Minimum 5-set package - If you're ready to start the construction process, this 5-set package is the minimum number of blueprint sets you will need. It will require keeping close track of each set so they can be used by multiple subcontractors and tradespeople.

The Standard 8-set package - For best results in terms of cost, schedule and quality of construction, we recommend you order eight (or more) sets of blueprints. Besides one set for yourself, additional sets of blueprints will be required by your mortgage lender, local building department, general contractor and all subcontractors working on foundation, electrical, plumbing, heating/air conditioning, carpentry work, etc.

Reproducible Masters - If you wish to make some minor design changes, you'll want to order reproducible masters. These drawings contain the same information as the blueprints but are printed on erasable and reproducible paper. This will allow your builder or a local design professional to make the necessary drawing changes without the major expense of redrawing the plans. This package also allows you to print as many copies of the modified plans as you need.

Mirror Reverse Sets - Plans can be printed in mirror reverse. These plans are useful when the house would fit your site better if all the rooms were on the opposite side than shown. They are simply a mirror image of the original drawings causing the lettering and dimensions to read backwards. Therefore, when ordering mirror reverse drawings, you must purchase at least one set of right reading plans.